Lecture Notes in Computer Science 12456

More information about this subseries at http://www.springer.com/series/7407

Tomáš Kozubek · Peter Arbenz ·
Jiří Jaroš · Lubomír Říha ·
Jakub Šístek · Petr Tichý (Eds.)

High Performance Computing in Science and Engineering

4th International Conference, HPCSE 2019
Karolinka, Czech Republic, May 20–23, 2019
Revised Selected Papers

 Springer

Editors
Tomáš Kozubek 🆔
VSB - Technical University of Ostrava
Ostrava-Poruba, Czech Republic

Jiří Jaroš 🆔
Brno University of Technology
Brno, Czech Republic

Jakub Šístek 🆔
Institute of Mathematics of the CAS
Prague, Czech Republic

Peter Arbenz 🆔
ETH Zurich
Zurich, Switzerland

Lubomír Říha 🆔
VSB - Technical University of Ostrava
Ostrava-Poruba, Czech Republic

Petr Tichý 🆔
Charles University
Prague, Czech Republic

ISSN 0302-9743   ISSN 1611-3349 (electronic)
Lecture Notes in Computer Science
ISBN 978-3-030-67076-4   ISBN 978-3-030-67077-1 (eBook)
https://doi.org/10.1007/978-3-030-67077-1

LNCS Sublibrary: SL1 – Theoretical Computer Science and General Issues

This Springer imprint is published by the registered company Springer Nature Switzerland AG
The registered company address is: Gewerbestrasse 11, 6330 Cham, Switzerland

Preface

This volume comprises the proceedings of the 4th International Conference on High Performance Computing in Science and Engineering – HPCSE 2019, which was held in the Hotel Soláň in the heart of the Beskydy Mountains, Czech Republic, on May 20–23, 2019. The biennial conference was organized by the IT4Innovations National Supercomputing Center at VSB - Technical University of Ostrava with the support of partner institutions: Brno University of Technology, Charles University, Czech Technical University in Prague, Institute of Geonics and Institute of Mathematics of the CAS, and Czech Network for Mathematics in Industry eu-maths-in.cz. The aim was to bring together specialists in high performance computing from fields such as applied mathematics, numerical methods, and parallel computing, to exchange experience and to initiate new research collaborations. We are glad that our invitation was accepted by distinguished experts from world-leading research institutions.

This conference has become an international forum for exchanging ideas among researchers involved in scientific and parallel computing, including theory and applications, as well as applied and computational mathematics. The focus of HPCSE 2019 was on models, algorithms, and software tools that facilitate efficient and convenient utilization of modern parallel and distributed computing architectures, as well as on large-scale applications.

The members of the Scientific Committee of HPCSE 2019 were Radim Blaheta, Martin Čermák, Zdeněk Dostál, Jaroslav Hron, Jiří Jaroš, Tomáš Kozubek, Jaroslav Kruis, Tomáš Oberhuber, Ivan Šimeček, Jakub Šístek, Petr Tichý, and Vít Vondrák.

The invited plenary talks were presented by:

- Andrea Bartolini (University of Bologna),
- Steffen Börm (Christian-Albrechts-Universität zu Kiel),
- Ben T. Cox (University College London),
- Dominik Göddeke (University of Stuttgart),
- Frédéric Hecht (Université Pierre et Marie Curie),
- Jakub Kurzak (University of Tennessee),
- Günther Of (Technische Universität Graz),
- Daniela di Serafino (University of Campania Luigi Vanvitelli),
- Garth Wells (University of Cambridge),
- Stefano Zampini (KAUST).

We gratefully acknowledge the support of the Ministry of Education, Youth and Sports from the National Programme of Sustainability (NPU II) through project "IT4Innovations excellence in science – LQ1602".

The HPCSE 2019 conference was a fruitful event, providing interesting lectures, showcasing new ideas, demonstrating the beauty of applied mathematics, presenting numerical linear algebra, optimization methods, and high performance computing, and starting or strengthening collaborations and friendships.

This meeting attracted about 100 participants from 10 countries. All participants were invited to submit an original paper to this book of proceedings. We give thanks for all contributions as well as for the work of the reviewers, and hope that this volume will be useful for readers. The proceedings were edited by Tomáš Kozubek, Peter Arbenz, Jiří Jaroš, Lubomír Říha, Jakub Šístek, and Petr Tichý.

Finally, we would like to cordially invite readers to participate in the next HPCSE conference, which is planned to be held at the same place on May 17–20, 2021.

On behalf of the organizers, Tomáš Kozubek.

Tomáš Kozubek
Peter Arbenz
Jiří Jaroš
Lubomír Říha
Jakub Šístek
Petr Tichý

Organization

Conference Chair

Tomáš Kozubek — VSB - Technical University of Ostrava, Czech Republic

Scientific Committee

Radim Blaheta	Institute of Geonics of the CAS, Czech Republic
Martin Čermák	VSB - Technical University of Ostrava, Czech Republic
Zdeněk Dostál	VSB - Technical University of Ostrava, Czech Republic
Jaroslav Hron	Charles University, Czech Republic
Jiří Jaroš	Brno University of Technology, Czech Republic
Tomáš Kozubek	VSB - Technical University of Ostrava, Czech Republic
Jaroslav Kruis	Czech Technical University in Prague, Czech Republic
Tomáš Oberhuber	Czech Technical University in Prague, Czech Republic
Ivan Šimeček	Czech Technical University in Prague, Czech Republic
Jakub Šístek	Institute of Mathematics of the CAS, Czech Republic
Petr Tichý	Charles University, Czech Republic
Vít Vondrák	VSB - Technical University of Ostrava, Czech Republic

Contents

Thermal Characterization of a Tier0 Datacenter Room in Normal and Thermal Emergency Conditions

Mohsen Seyedkazemi Ardebili[1]([✉])[iD], Carlo Cavazzoni[2][iD], Luca Benini[1,3][iD], and Andrea Bartolini[1][iD]

[1] Universitá degli Studi di Bologna, Viale Risorgimento, 2, 40136 Bologna, Italy
{mohsen.seyedkazemi,luca.benini,a.bartolini}@unibo.it
[2] CINECA, Via Magnanelli 6/3, Casalecchio di Reno, 40033 Bologna, Italy
c.cavazzoni@cineca.it
[3] Eidgenössische Technische Hochschule Zürich, Gloriastrasse 35, 8092 Zürich, Switzerland
lbenini@iis.ee.ethz.ch
https://ee.ethz.ch/
https://www.dei.unibo.it/
https://www.cineca.it/

Abstract. Datacenters are at the heart of the AI, Industry 4.0 and cloud revolution. A datacenter contains a large number of computing nodes hosted in a large temperature-controlled room. Due to the increasing total power and power density of computing nodes, the overall datacenter compute capacity is often capped by peak power consumption and temperature bottlenecks. To preserve the homogeneous performance assumption between all the nodes, complex cooling solution are required, but they might not be sufficient. In this work, we analysed and characterised the thermal properties of a Tier0 datacenter deploying advanced hybrid cooling technologies: specifically, we studied the spatial and temporal heterogeneity during production and cooling emergency hazards. This paper gives first quantitative evidence of thermal bottlenecks in real-life production workload, showing the presence of significant spatial thermal heterogeneity which could be exploited by thermal-aware job scheduling and datacenter-room run-time workload adaptation and distribution.

Keywords: HPC · Thermal characterization · Power consumption

1 Introduction and Related Works

With the growth of the computing demand from a broad set of societal, industrial and science application, datacenter have become a key component of the whole

This work has been partially supported by the EU H2020 ICT/2018 project IoTwins (g.a. 857191).

T. Kozubek et al. (Eds.): HPCSE 2019, LNCS 12456, pp. 1–16, 2021.
https://doi.org/10.1007/978-3-030-67077-1_1

ICT. In the US a single dollar invested in HPC generates in average 43\$ of profit while in Europe the Return on Investment (ROI) of each euro invested in HPC generates in average 69€ in profit and 867€ of increased revenues [9].

A datacenter is composed of several computing rooms, each hosting several racks containing tens/hundreds of computing nodes. The power consumption of these installations ranges from few to tens of MWatts. Additional power is required to remove the heat generated by the active electronics. Summit [8] which is today the most powerful supercomputer worldwide consumes 11 MWatts for the computation and an additional 1.32 MWatts for the cooling. To achieving this cooling efficiency Summit adopts a sophisticated computing node design and hot water cooling solution [16]. Today's datacenters of Google achieve similar cooling efficiency and pay an equivalent additional 12% of power consumption for power delivery and cooling dissipation [11].

Traditional cooling methods, based on computer room air conditioners (CRAC), or computer room air handlers (CRAH) have been enhanced with free-cooling mode, i.e., the capability to exploit the outside air, using only the AC blowers to circulate it in the room [15]. Rear Door Heat Exchangers (RDHX) are used to augment the computing density in air-cooled computing rooms. While to further reduce the cooling, hot water cooling is used to remove heat [16,18]. Hot water cooling requires special and costly heat exchanger (cold plate) to reduce the thermal resistance and in general lead to a higher silicon temperature in the computing unit than cold water cooling [3,5].

Several works in literature have analyzed the impact of heat dissipation in datacenter components.

The first set of works focus on the chip-level thermal effects and show that at chip-level exists hotspots and significant thermal gradients which can be exploited for improving core's performance and energy-efficiency [3,4,6,7,12]. Druzhinin et al. have studied the impact of the coolant temperature increase in a datacenter blade with hot water cooling. The authors show that an increment of 40 °C in the coolant causes 20% of additional leakage power and a consequent decrease in the performance of 0.5% [14].

The second set of works focus on the machine level [10,13,18]. These works characterize the effect of performance variability between nominally equal computing nodes. Marathe et al. [13] show that in power-constrained computing nodes, the hardware control logic turns the process variation effects into a performance and core's frequency variation. This can lead to significant application time-to-solution overheads in parallel applications. While all the previously mentioned papers highlight and characterize the side-effects of inlet temperature, performance variation, energy efficiency and parallel job performance in a datacenter, none of them has characterized the temperature variation in a datacenter's room.

In this work, we characterize the temperature distribution of a Tier0 datacenter hosting the Marconi supercomputer [8,17], which is ranked 21st in the list of the most powerful supercomputer worldwide and features hybrid and free-cooling technologies. To carry out the analysis, we have collected the entire

Marconi node's telemetry data for a month of activity. During the selected period (01.06.2019–01.07.2019), the ambient temperature has ranged from 12 °C to 38 °C. Our analysis shows that:

- The inlet temperature of the nodes increases vertically. With an average difference of 6.5 °C from the top and bottom nodes. Moreover, the bottom nodes face a higher variability of the inlet temperature than the top nodes in the rack as an effect of a stronger dependency of their inlet air with the CRAC outlet temperature. This is less strong with top nodes in the rack due to a stronger dependency of their inlet air from heat re-circulation.
- The inlet temperature significantly changes in the floorplan. We measured up to 10.8 °C difference for the monthly average chassis temperature for chassis at the same height in the racks. Interestingly the monthly average hotspot position in the floorplan is correlated with the chassis height.
- In the observed period, the datacenter faced a thermal hazard which has compromised the liquid cooling capacity of the room (used by the RDHX). We carefully analyze room temperature during this rare but extremely critical event. Our measurement shows that the effect of the thermal emergency caused an increase in the average temperature of the computing nodes with a modification of the hotspot location.

Our results show that the inlet temperature in a datacenter is heterogeneous and significant patterns are stable for long periods and visible in monthly average. If accurately modelled this information can be used to improve job scheduling and improve the datacenter's cooling efficiency.

In Sect. 2, we present the methodology used to conduct the analysis. In Sect. 3, we present the results of our analysis.

2 Methodology

We focus our study on the MARCONI Tier-0 cluster in the CINECA datacenter, which is based on the Lenovo NeXtScale platform. MARCONI is composed of two partitions, one (A3) based on Intel Skylake processors ($8PFlops$) and the other (A2) based on Intel KnightLandings (KNL) processors ($11PFlops$). We focused our analysis on the room hosting the largest partition of MARCONI (A2), namely the Marconi KNL room. The MARCONI A2 cluster is based on the 68-cores Intel Xeon Phi7250 (KnightLandings) at 1.4 GHz, with many-core architecture (Intel OmniPath Cluster), provided about 250 thousand cores (68 cores/node, 244.800 cores in total) with the computational power of around 11Pflop/s. Each node has 16 GB/node MCDRAM + 96 GB/node DDR4 [17]. The CINECA datacenter features a holistic monitoring framework, namely ExaMon [2], which aggregates a wide set of telemetry data.

Figure 1 depicts the layout of the HPC Marconi KNL room in CINECA. In the Marconi KNL room, 46 + 1 racks (one of them is a rack of switches) are located in three rows. Each rack is composed of 18 chassis in different height, and each chassis has four computing nodes. Chassis one (C1) is in the bottom, and

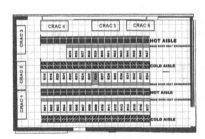
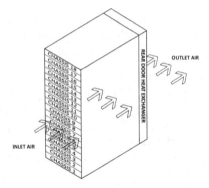

(a) Marconi KNL Room in CINECA Datacenter.

(b) Racks Arrangements of Marconi KNL Room in CINECA Datacenter.

Fig. 1. Racks arrangements of Marconi KNL room in CINECA datacenter.

chassis 18 (C18) is the highest one. There are six computer room air conditioning (CRAC) units which support the two cold aisles. The RDHX of racks are in the hot aisle. For each node and its associated components, such as voltage regulators and fans, the Intelligent Platform Management Interface (IPMI) provide remote telemetry access to the built-in sensors [1]. The ExaMon monitoring system collects sensor data with the IPMI interface with 20 s sampling rate [2]. ExaMon monitored data is stored in its internal KairosDB database as time traces and remotely accessible through RESTfull APIs [2].

We focus our analysis on the following metrics: (i) inlet temperature (BB_Inlet_Temp) which senses the temperature of a node close to the cold aisle; (ii) outlet temperature (Exit_Air_Temp), which senses the temperature of a node close to the RDHX and the hot corridor. (iii) The node power, which is derived from the power measured for the two power supplies of each chassis (namely PS1_Input_Power and PS2_Input_Power metrics). The power consumption and workload are related, and this is not the focus of the manuscript. All these metrics are available in ExaMon. This study investigates the spatial and temporal heterogeneity during production and cooling emergency hazards for the period from 2019-06-01 00:00:00 to 2019-07-01 00:00:00 over the 3312 Marconi KNL nodes. To conduct the study, we used the following methodology. We extracted the data by using the RESTfull API provided by ExaMon. We use Python programming language scripts. With Examon-KairosDB, we build suitable datasets, and with Python codes, we performed data analysis and plots [2]. Table 1 summarizes the characteristics of the dataset used in this study, and the boxplot in Fig. 2 shows the shape of the distribution of inlet and outlet temperatures, and power consumption of nodes in June 2019. To generate these boxplots, we use all the collected data of sensors during June 2019. As it be noticed, there is no overlap in the interquartile range between the inlet and outlet temperatures.

Table 1. Characteristics of dataset

Name of parameter	Value
Number of racks	46
Number of chassis Per Rack	18
Number of nodes	3312
Number of metrics	42 IPMI with knl tag
Sampling rate	20 s
Period of study	From 2019-06-01 00:00:00 to 2019-07-01 00:00:00
Thermal emergency	2019-06-28

3   Experimental Results

To study the thermal characteristic of the Marconi KNL room, we investigated the spatial and temporal aspects of temperature and power consumption of nodes in the room during June 2019. This study contributes the 3D view of the thermal and power characteristics of the room by utilizing the heat-map of distribution of the power consumption and temperature of nodes, and also, different chassis-level analysis that represents the power consumption and thermal variation in different height of the room.

Subsection 3.1 analyzes the static spatial gradients present in the computing room. The analysis is conducted by averaging each metrics for the entire month and studying their correlation and spatial variation. Differently, Subsect. 3.2 analyzes the temporal variations by computing the average and the min-max variation on a per-day base for the entire computing room. Finally, Subsect. 3.3 focuses on the day for which the computing room has faced a rare cooling hazard.

3.1   Spatial Study

Figure 3 shows the boxplot of inlet and outlet temperatures. In the x-axis, the chassis number: higher vertical position is represented by bigger chassis number, being C1 the bottom one, and C18 the top chassis in a rack. The y-axis shows the temperature in C. For each chassis number, we collected the temperature of nodes located in the different racks in the room still in the same chassis-number and as a consequence in the same height.

For each chassis number, the boxplot generated among all the nodes belonging to a given chassis number in the room (4 nodes per chassis $\times$ 46 racks) and all the samples (23.8 M samples) collected in June 2019. Figure 4 reports the boxplot of power consumption of chassis in Watt. This plot generated in the same approach as Fig. 3.

Figure 3 demonstrates the presence of a vertical spatial thermal gradient. The nodes which hosted in the chassis-2 of racks on average have a minimum inlet,

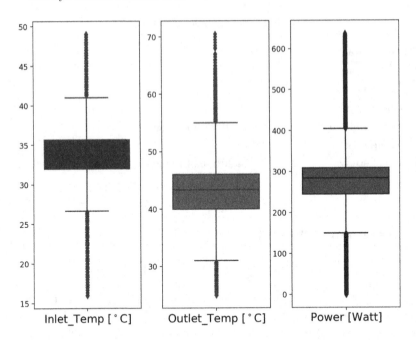

Fig. 2. Boxplot of inlet and outlet temperatures and power consumption of computing nodes in June 2019.

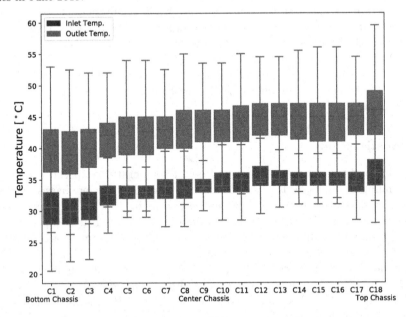

Fig. 3. Boxplot of inlet and outlet temperature of computing nodes in different chassis in June 2019.

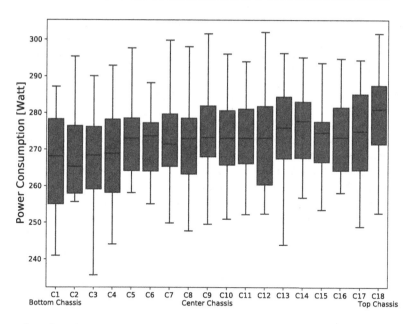

Fig. 4. Boxplot of power consumption of computing nodes in different chassis in June 2019.

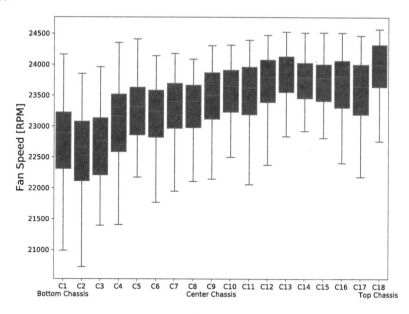

Fig. 5. Boxplot of fan speed (RPM) of computing nodes in different chassis in June 2019.

and outlet air temperature, therefore, these nodes on average are coldest nodes in the room (∼6 °C colder than chassis-18 ones).

Figure 5 illustrates the distribution of fans speed in different chassis-numbers. Measured data confirm that fans of nodes of chassis-2 work with lower speed/RPM and consume 15.8 W less (∼6%) than nodes of chassis-18.

We then studied the thermal variation in different chassis-level by averaging for the entire June the daily variation for the different analyzed metrics for different chassis-levels. Figure 6 reports these values and indicates that chassis-2 endured maximum thermal variation, and it has experienced on average 7.3 °C thermal variations in inlet temperature and 12.1 °C in the outlet. From the plot, we can notice that the thermal variation drops vertically and is more severe for the inlet temperature. This effects can be explained by the fact that the inlet temperature of the lower chassis is closer to the CRAC outlet air which, due to free-cooling follow the external ambient temperature and daily variations. The inlet air temperature for nodes in the higher chassis is instead affected also by the rack dissipated heat as the effect of heat recirculation. The lower variation for the outlet air w.r.t. the inlet air can be explained by the larger fan speed of the nodes in the higher chassis.

We report two heat-map plots of Marconi KNL room that illustrate the distribution of the average inlet temperature in different racks at two different heights (bottom and top of the racks) of the room in June. The bar-colour shows the temperate in C degree. The top plot in Fig. 7 describes the inlet thermal status of nodes in the top of the room (chassis-18), and the bottom plot in the same figure shows inlet thermal status at the bottom of the room (chassis-2). For both the plots, the center row of racks is at the colder temperature. In average in June

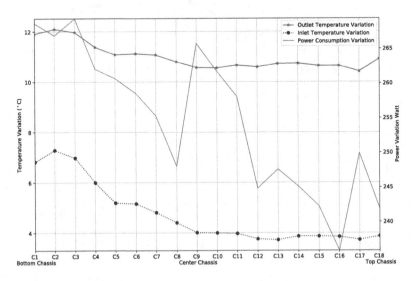

Fig. 6. Average inlet and outlet temperature variation and power consumption variation of computing nodes in chassis in June 2019.

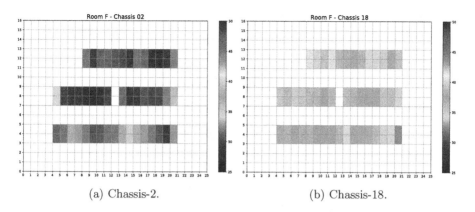

(a) Chassis-2.   (b) Chassis-18.

Fig. 7. Average inlet heat map of Marconi KNL room on June 2019.

the bottom nodes of each rack (chassis-2) had maximum 34.35 °C and minimum 25.46 °C as inlet air temperature, for the top nodes of each rack (chassis-18) that was 42.1 °C, and 31. °C respectively. Moreover, for the same height just by moving in a horizontal/plane direction for the top of the racks (chassis-18 height), the room had 10.8 °C of thermal variation, which is notable inlet temperature heterogeneity. At the bottom of the racks (chassis-2 height), the room had 2 °C lower thermal variation. It must also be noted that the two horizontal sections of the average room temperature show different hotspots locations. This result suggests that the horizontal heat distribution vary vertically in the room. This effect poses challenges in proactive room level thermal management, as all the 3D thermal maps should be considered for optimizations.

Figure 8 reports the average power consumption of different racks in the room in June. The bar-color shows electrical power in KWatt. In the computing room, the maximum average power consumption of a rack was 20.4 KWatt, the minimum of 14.0 KWatt with a standard deviation of 1.8 KWatt. The power consumption correlated to the inlet temperature with a correlation coefficient (CC) equal to 0.68. The outlet temperature correlated with the inlet temperature with CC = 0.91. Finally, the outlet temperature correlated with the power consumption with CC = 0.88. We can conclude that there is an intertwined dependency between node's power consumption, inlet temperature and outlet temperature, which can be exploited for optimizing the room cooling and saving cooling energy.

3.2   Temporal Study

In this subsection, we analyze the temporal variations in the heat dissipation of the datacenter room.

Figure 9 shows in the x-axis the days of June 2019, and y-axis reports the average inlet and outlet temperature of nodes in the room for each day in C degree. Each reported value corresponds to a day and is computed as the average

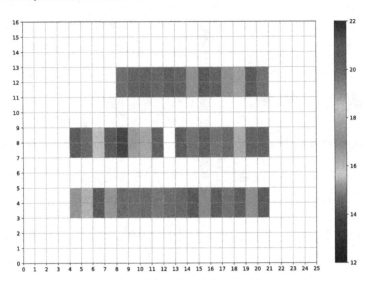

Fig. 8. Average power consumption [KWatt] of racks of Marconi KNL room in June 2019.

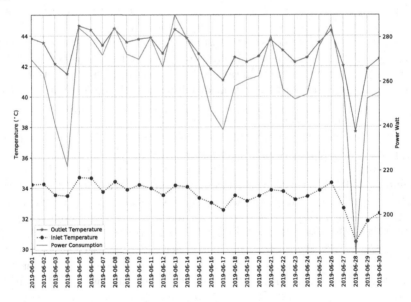

Fig. 9. Average inlet and outlet temperature and power consumption of computing nodes in different days of June 2019.

among all the nodes in the room (4 nodes per chassis × 18 chassis × 46 racks) and all the samples in a day (3 samples per minutes × 1440 min per day).

From the plot, we can notice that all the reported metrics were relatively constant for all the days of June except the 28th which had a thermal capacity failure. In addition, both the outlet and inlet temperature daily variation follows the average node power consumption. It must be noted that even if during the thermal emergency the outlet and inlet temperature increased due to the compromised RDHX cooling capacity their average value in the day is lower due to the counteracting action taken by the machine administrators that reduced the room power consumption as we will see in the next subsection. Indeed, the node average power consumption was 280 W on the 11th of June and 183 W on the 28th of June. Although with around 100 W reduction in the power consumption of each node the inlet and outlet temperatures of nodes decreased, its thermal variation dramatically raised (Fig. 10).

Measurement reveals that nodes in the thermal emergency day on average had 14.7 °C thermal fluctuations in the inlet and 23.7 °C in the outlet temperatures. It must be noted that during regular days, the average thermal fluctuation (computed as the average of the daily min-max variation) is lower for the inlet temperature than the outlet temperature.

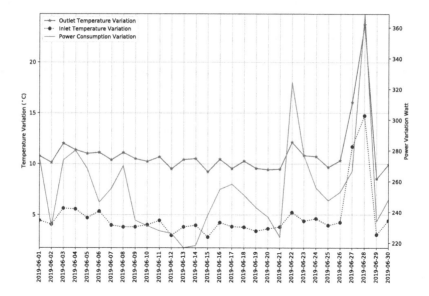

Fig. 10. Average inlet and outlet temperature variation and power consumption variation of computing nodes in different days of June 2019.

3.3   Thermal Emergency

In this section, we analyze the heat variation of the datacenter room during the thermal emergency day (28th of June 2019). Figure 11 shows in x-axis time, and left and right y-axis respectively shows the temperature in C degree and power consumption in Watt. In Fig. 11, we report the inlet, outlet temperatures, and power consumption of a node during the thermal hazard day and as it can be seen the thermal hazard starts after 16:00 o'clock then it reaches its' peak around 17:20 o'clock. In this period, the power consumption decreased to zero, which means the computing nodes were turned off for a while, so, there is loss-data for the outlet temperature. Therefore we extracted the data of five snapshots in time corresponding to before, during and after the thermal emergency. The 10:00, 12:00, and 16:00 o'clock snapshots corresponds to the node's condition before the peak of thermal emergency. The 17:20 and 19:00 o'clock snapshots provide information for the peak and after the peak of the thermal emergency. Finally, the 21:00 o'clock snapshot corresponds to the recovery after the thermal emergency.

Figure 12 shows in the x-axis the chassis number and in y-axis the inlet temperate in C degree. As can be noted from Fig. 12, although generally inlet temperature increase with height, the thermal pattern of chassis in the thermal emergency period is quite different from the one during the normal conditions. Around the hazard at 17:20 and 19:00, (i) the inlet temperature increases of ∼5 °C (ii) the hotspot from chassis-18 moved to the chassis-17 and 15 also, (iii) the global minimum for temperature was not on the chassis-2, and (iv) the chassis-1 and 4 were colder than the chassis-2. We reported the outlet temperature in Fig. 13, which shows in the x-axis the chassis number and in the y-axis the outlet temperate in C degree. We can notice that during thermal hazard, the outlet temperature was colder than during a typical day. This outlet temperature reduction was more prominent for the higher chassis than the lower one. For example, chassis-15 faced a 10 °C temperature reduction.

An explanation of this effect can be found in Fig. 14, which provides the average power consumption for the nodes in different chassis during the five time-snapshot examined for the thermal emergency day. The x-axis of Fig. 14 shows the chassis number, and the y-axis shows the average power consumption in Watt. We find that the reduction in outlet temperature correlates with a sharp decrease in power consumption from 270 W to 6 W, as it is evident in Fig. 14. Moreover after the thermal emergency at the 21:00 o'clock the average node power has been reduced to only 150 W. This power reduction was due to the machine administrator intervention, which initially switched off the nodes (17:20) and then started to bring up the nodes gradually.

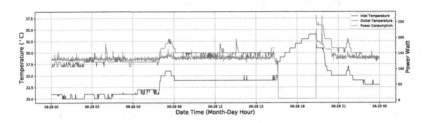

Fig. 11. Inlet, outlet temperature and power consumption of a computing node on 28 June 2019 the day of thermal emergency.

To finalize the study of the thermal emergency day in Fig. 15 we report two heat-map plots of Marconi KNL room that show the distribution of the inlet temperature in different racks at two different heights (bottom and top of the racks) of the room at 17:20 on the 28th of June 2019. The bar-colour shows the temperature in C degree. From the figure, we can notice that the centre row is almost always colder than others like the normal. At 17:20 the bottom nodes of each rack (chassis-2) had maximum 37 °C and minimum 30 °C as inlet air temperature, for the top nodes of each rack (chassis-18) that was 43 °C, and 36 °C respectively. Moreover, for the same height just by moving in a horizontal/plane direction for the top of the racks (chassis-18 height), the room had 7 °C of thermal variation, which is lower than the normal condition 10.8 °C (Fig. 7) and also in the bottom we have the same amount of variation 7 °C. Therefore the thermal heterogeneity of room was reduced.

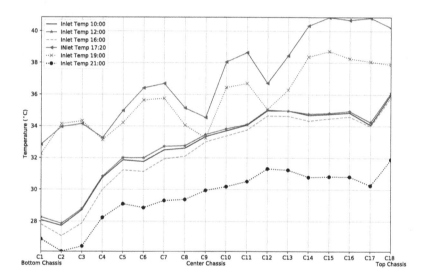

Fig. 12. Average inlet temperature of computing nodes in different chassis in different time instances on 28 June 2019 the day of thermal emergency.

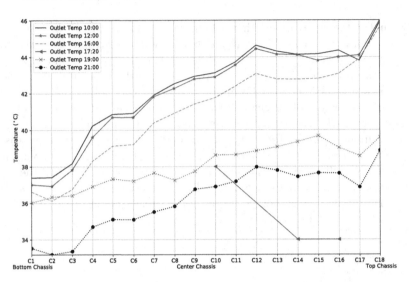

Fig. 13. Average outlet temperature of computing nodes in different chassis in different time instances on 28 June 2019 the day of thermal emergency.

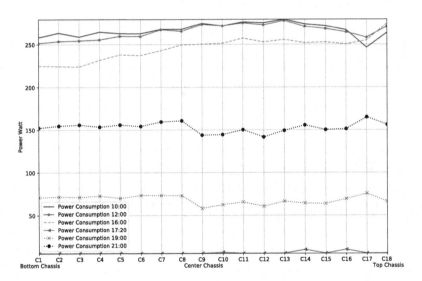

Fig. 14. Average power consumption of computing nodes in different chassis in different time instances on 28 June 2019 the day of thermal emergency.

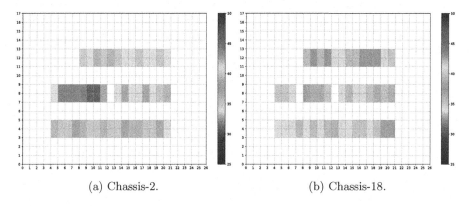

(a) Chassis-2.   (b) Chassis-18.

Fig. 15. Heatmap of Marconi KNL room on 28 June 2019 at 17:20 the day of thermal emergency.

4   Conclusion

In this work, we analyzed the spatial and thermal heat dissipation characteristics of the CINECA Marconi KNL room. The study revealed that nodes hosted in the top chassis of racks are in worse thermal conditions than bottom nodes. This has a direct impact on the average power consumption of the nodes, which is higher for the top nodes. These nodes can consume up to 6% more power due to a higher fan speed than bottom nodes. The study of the thermal map revealed that the center row of racks in the Marconi KNL room is colder than the other two rows; overall, this was valid for normal and thermal hazard condition. The hotspot varies vertically as well as during the thermal emergency condition. We can conclude that the study of the spatial and thermal heat dissipation characteristics revealed significant non-idealities and heterogeneity which if modelled, can be leveraged by thermal-aware job-scheduler and room-level power management run-times.

References

1. Intel Server Board S2600IP and Workstation Board W2600CR Technical Product Specification (2013)
2. Bartolini, A., et al.: Paving the way toward energy-aware and automated datacentre. In: Proceedings of the 48th International Conference on Parallel Processing: Workshops, ICPP 2019, pp. 8:1–8:8. ACM, New York (2019)
3. Bartolini, A., Conficoni, C., Diversi, R., Tilli, A., Benini, L.: Multiscale thermal management of computing systems-the MULTITHERMAN approach. IFAC PapersOnLine **50**(1), 6709–6716 (2017)
4. Beneventi, F., Bartolini, A., Cavazzoni, C., Benini, L.: Cooling-aware node-level task allocation for next-generation green hpc systems. In: Proceedings of the 2016 International Conference on High Performance Computing and Simulation (HPCS), pp. 690–696. IEEE (2016)

5. Conficoni, C., Bartolini, A., Tilli, A., Cavazzoni, C., Benini, L.: Integrated energy-aware management of supercomputer hybrid cooling systems. IEEE Trans. Ind. Inform. **12**(4), 1299–1311 (2016)
6. Coskun, A.K., Ayala, J.L., Atienza, D., Rosing, T.S.: Modeling and dynamic management of 3D multicore systems with liquid cooling. In: Proceedings of the 17th IFIP International Conference on Very Large Scale Integration (VLSI-SoC 2009), pp. 35–40. IEEE (2009)
7. Diversi, R., Tilli, A., Bartolini, A., Beneventi, F., Benini, L.: Bias-compensated least squares identification of distributed thermal models for many-core systems-on-chip. IEEE Trans. Circ. Syst. I Regul. Pap. **61**(9), 2663–2676 (2014)
8. Dongarra, J.J., Meuer, H.W., Strohmaier, E.: 29th top500 supercomputer sites. Technical report, Top500.org (1994)
9. ETP4HPC: Strategic research agenda (2017)
10. Fraternali, F., Bartolini, A., Cavazzoni, C., Tecchiolli, G., Benini, L.: Quantifying the impact of variability on the energy efficiency for a next-generation ultra-green supercomputer. In: Proceedings of the 2014 International Symposium on Low Power Electronics and Design, pp. 295–298. ACM (2014)
11. Gao, J., Jamidar, R.: Machine learning applications for data center optimization (2014)
12. Kim, J., Sabry, M.M., Ruggiero, M., Atienza, D.: Power-thermal modeling and control of energy-efficient servers and datacenters. In: Khan, S.U., Zomaya, A.Y. (eds.) Handbook on Data Centers, pp. 857–913. Springer, New York (2015). https://doi.org/10.1007/978-1-4939-2092-1_29
13. Marathe, A., Zhang, Y., Blanks, G., Kumbhare, N., Abdulla, G., Rountree, B.: An empirical survey of performance and energy efficiency variation on intel processors. In: Proceedings of the 5th International Workshop on Energy Efficient Supercomputing, E2SC 2017, Denver, CO, USA, pp. 9:1–9:8. ACM, New York (2017)
14. Moskovsky, A., Druzhinin, E., Shmelev, A., Mironov, V., Semin, A.: Server level liquid cooling: do higher system temperatures improve energy efficiency? Int. J. Supercomput. Front. Innov. **3**(1), 67–74 (2016)
15. Pore, M., Abbasi, Z., Gupta, S.K.S., Varsamopoulos, G.: Techniques to achieve energy proportionality in data centers: a survey. In: Khan, S.U., Zomaya, A.Y. (eds.) Handbook on Data Centers, pp. 109–162. Springer, New York (2015). https://doi.org/10.1007/978-1-4939-2092-1_4
16. Rogers, J.: ORNL's warm water HPC facilities and control systems (2019)
17. Rossi, E.: MARCONI-A2 (KNL) (2017)
18. Shoukourian, H., Wilde, T., Huber, H., Bode, A.: Analysis of the efficiency characteristics of the first high-temperature direct liquid cooled petascale supercomputer and its cooling infrastructure. J. Parallel Distrib. Comput. **107**, 87–100 (2017)

Towards Local-Failure Local-Recovery in PDE Frameworks: The Case of Linear Solvers

Mirco Altenbernd[1], Nils-Arne Dreier[2], Christian Engwer[2], and Dominik Göddeke[1,3(✉)]

[1] University of Stuttgart, Allmandring 5b, 70569 Stuttgart, Germany
`dominik.goeddeke@mathematik.uni-stuttgart.de`
[2] University of Münster, Orleansring 10, 48149 Münster, Germany
[3] Stuttgart Center for Simulation Science (SimTech), Stuttgart, Germany

Abstract. It is expected that with the appearance of exascale supercomputers the mean time between failure in supercomputers will decrease. Classical checkpoint-restart approaches are too expensive at that scale. Local-failure local-recovery (LFLR) strategies are an option that promises to leverage the costs, but actually implementing it into any sufficiently large simulation environment is a challenging task. In this paper we discuss how LFLR methods can be incorporated in a PDE framework, focussing at the linear solvers as the innermost component. We discuss how Krylov solvers can be modified to support LFLR, and present numerical tests. We exemplify our approach by reporting on the implementation of these features in the DUNE framework, present C++ software abstractions, which simplify the incorporation of LFLR techniques and show how we use these in our solver library. To reduce the memory costs of full remote backups, we further investigate the benefits of lossy compression and in-memory checkpointing.

Keywords: PDE frameworks · Fault tolerance · Lossy compression

1 Introduction

Exascale supercomputers are expected to appear in the near future and pre-exascale systems are already operational [35]. Early after reaching petascale, the power, concurrency, locality, resilience and scalability challenges have been identified [20,21,30,41], and a recent PhD thesis by Nielsen [36] summarizes the progress that has been made on the road towards exascale as of 2018.

1.1 Motivation

In this work, we focus on the resilience challenge, as emphasised by a number of potentially alarming reports: The mean time between failure (MTBF) on the Titan supercomputer (#1 on the TOP500 in November 2012, decommissioned

© Springer Nature Switzerland AG 2021
T. Kozubek et al. (Eds.): HPCSE 2019, LNCS 12456, pp. 17–38, 2021.
https://doi.org/10.1007/978-3-030-67077-1_2

in 2019) has been in the range of hours [24], and similar but slightly lower values have been reported for the Blue Waters system [18] and Tianhe-2 [16], the #1 supercomputer from 2013–2015. The architectures of these exemplary machines differ vastly, indicating that progress towards a 'hardware solution' to the resilience challenge proceeds at a much slower rate than solutions towards other exascale challenges, e.g., as in the continued establishment of 'MPI+X' programming models. This is further exemplified in detailed studies of failures in HPC systems, see for instance [9, 26, 38], and by surveying the expectations spanning the last decade [14, 15, 19].

It is a known or at least a widely accepted fact that global, synchronous checkpoint-restart strategies are already prohibitively expensive at scale, and continue to become worse. For instance, 2000 s for a 64 TB dump on 1000 nodes has been reported already in 2014 [40]. However, such strategies are, if at all, the ones that are actually implemented in large-scale software, leading to the unfavorable situation that the effects of a lack of 'better' solutions combined with the slow improvement in hardware and middleware exponentiate themselves.

Already quite early in the pursuit of the exascale challenges, so-called 'local-failure local-recovery (LFLR)' strategies have been suggested, see e.g. [39, 43]. The LFLR philosophy has the potential to change fault mitigation from a reactive to a proactive technique: If the time to compute and store a checkpoint is minimized and is ideally almost completely hidden from the progress of the application, then the checkpointing frequency can be increased compared to conventional approaches, which immediately implies reduced recovery times in case of a failure, because the 'distance' from the corrupted to the sane data is reduced.

While LFLR is undoubtedly the prime choice, actually implementing it into any sufficiently large simulation environment is challenging. We focus on simulation frameworks for partial differential equation models (PDEs), that are modularly designed not for a single application but for a whole range of different ones. In fact, it means that almost all components need to be adapted, in contrast to global checkpoints, and manual restarting using the queueing system. We illustrate the complexity with a node loss scenario: When a local node dies, by definition of the MPI standard all local parts of the distributed data are lost. With ULFM (a middleware solution, see Subsect. 1.3), all MPI communicators can be restored, for instance by swapping in spare processes, but the resulting ranks are then 'empty'. To enable the application to proceed, all local state has to be restored. For PDE frameworks, this means that the local part of the grid has to be reconstructed, then the discretization has to be reassembled, and finally, the state of the (nonlinear) solver or even the inversion or optimization loop has to be restored, up to the point where the reconstructed state matches that of the non-lost ranks in the simulation. All these are typically implemented as separate modules, possibly developed by different parties in different sub-projects, and designed with interoperability e.g. between solvers for different discretizations on different grids in mind, necessitating sophisticated techniques for the 'catching up' of the lost rank(s).

1.2   Contribution

We argue to tackle proactive (and ultimately automatic) LFLR for PDE software frameworks in a bottom-up approach, i.e., for each module in a large framework separately. For this, we consider the Distributed Unified Numerics Environment (DUNE, see Subsect. 2.1), and mention that our ideas can also be implemented in other frameworks as well. In this paper focus on linear solvers as one core building block: Linear solvers are often the most time-consuming part of an application, and exhibit the degree of complexity necessary to develop more general techniques. Following the DUNE interoperability philosophy, we develop abstract interfaces that encapsulate all functionality needed for LFLR, and demonstrate their practical use only for linear solvers. In particular, we assume that all input data for the linear solvers, i.e., the locally lost parts of the matrix and the right hand side(s), can be treated by similar interfaces but different techniques under the hood, that we however have not implemented yet.

For linear solvers, it turns out that the local checkpointing and recovery procedure can potentially be substantially accelerated on large scale systems if checkpoints are stored in a compressed way. In this paper, we thus examine techniques to efficiently store (compressed) local checkpoints, in remote memory instead of on disk, and to use these checkpoints to locally recover the state of the global linear preconditioned Krylov solver in such a way that global convergence is only minimally affected by the recovery.

Our compression-based approach to checkpoint-restart can be useful even in the case of full checkpoints taken before the iterative solve, in a transition phase where not all modules support LFLR yet: In case of a node loss the processes can roll back to their full checkpoint followed by a 'solver recovery' which creates a good initial guess based on available solver checkpoints. In certain cases, this yields a superior recovery approach, as detailed throughout the paper.

1.3   Related Work

The current MPI standard mandates that all ranks terminate if one does. ULFM (User Level Failure Mitigation, [10,11]) is a solution to this problem, and will be integrated into the MPI4 standard. Some experience exists with PDE solvers on the application level [3], and for various in-situ recovery techniques with shrink and/or substitute [5], and for the connection to LFLR [43].

Similar ideas to our approach have been pursued: Cantwell et al. apply message logging and remote in-memory checkpointing to Nektar++ [13]. Transparent recovery via implicitly coordinated, diskless, application-driven in memory checkpointing has been added to fenix [23]. Also, the waLBerla framework contains a resilient, diskless, and distributed checkpointing scheme to regularly create snapshots of simulation data using ULFM [31]. Losada et al. [34] recently demonstrated that by combining ULFM, the ComPiler for Portable Checkpointing (CPPC) tool, and the Open MPI VProtocol system-level message logging to construct a local rollback mechanism, only failed processes need to be recovered from the last checkpoint, while consistency before further progress in the execution is achieved through a two-level message logging process.

Following the ideas of algorithm-based fault tolerance (ABFT, [28]), we have previously used the existing hierarchy of linear multigrid solvers to compress checkpoints [25] and to detect silent data corruption [4]. This has also involved the solution of local auxiliary Dirichlet problems to recover from a fault, see also Huber et al. [29], which we employ here too.

Data compression receives increased interest in scientific computing [17, 33,42]. Combining the ideas of ABFT and compression leads to so-called interpolation-restart (IR) strategies. Some examples include interpolating lost data from surviving nodes [32], IR for Krylov solvers [1] and for eigensolvers [2]. The idea we pursue in this paper lies somewhere in between.

2   Preliminaries

In this section, we outline the setting in which we implemented our approach, and discuss specific related work.

2.1   DUNE – The Distributed Unified Numerics Environment

DUNE [8] is a framework for the grid-based numerical solution of PDEs. It is being developed at several universities and research institutes, for more than 15 years under a free and open source license. The main goal of DUNE is the definition of interfaces for different parts of a PDE solver and specific implementations of these interfaces. Due to its modularity, it is highly flexible. The *core* modules consist of a grid interface [6,7], basic infrastructure, linear algebra, and finite element bases. The core modules are extended by grid modules, discretization modules, and several extensions and user modules.

Specifically we mention DUNE-COMMON and DUNE-ISTL. The DUNE-COMMON module provides general infrastructure and foundation classes. As part of this it defines a lightweight abstraction on top of MPI, which allows to introduce fault tolerance features without modifying the whole code base. We discuss this in more detail in the next subsection.

The module DUNE-ISTL (Iterative Solver Template Library, [12]) contains data types and methods for the solution of linear systems. Iterative solvers are implemented using dynamic polymorphism, which makes them configurable at runtime. Unlike other PDE frameworks, DUNE-ISTL does not rely on a global numbering of the unknowns, instead the building blocks (operator, preconditioner and scalar product) take care of the communication and consistency of the data. This makes DUNE well-suited for the design and implementation of LFLR strategies.

2.2   Restoring MPI Communicators with ULFM and Distributed Exceptions

In fault tolerance, the distinction between hard and soft errors is important, i.e. errors that do or do not terminate program execution. Current MPI standards can not handle hard errors like node losses. For the MPI-4 specification the ULFM proposal [10,11] suggests an approach, where MPI does not replace the failed rank, but provides users with sufficient information and functionality to react locally on all ranks to this failure.

From a framework point of view we aim at hiding any such details behind a high-level interface. This interface allows to react to possible changes in the proposal, but it also allows us to introduce appropriate abstractions, which help our integration into the existing code base. Our code is based on modern C++ and as such makes excessive use of exceptions for error management. This does usually not integrate well with MPI, as exceptions are only handled locally. The failing rank is then in a different state than all other ranks, which can immediately lead to dead-locks. In [22] we proposed an approach to introduce parallel exceptions into C++ code. In this paper we rely on an implementation of this proposal in DUNE. We extend the existing MPI abstractions in DUNE-COMMON, so that now all ranks receive a particular exception in case of soft or hard failures, allowing a synchronized reaction to mitigate failures locally.

Our implementation ensures that in case of a failure an exception is received on all surviving ranks. The term 'surviving' here means, that the rank is capable to continue the computation. We rely on two methods from the ULFM proposal: MPIX_Comm_revoke, which revokes the communicator for any communication and MPIX_Comm_agree to agree on the error state. Once a rank calls a communication method on a revoked communication an error is raised which is then mapped to an exception in our implementation. A working communicator can be recovered by calling MPIX_Comm_shrink and its siblings, on which the computation can be continued, after the error state has been resolved.

This functionality is typically not available in default MPI installations on clusters yet. To test our recovery strategies we have created a stand-alone library, which uses the MPI P-interface to equip any MPI implementation with the MPIX_Comm_revoke, MPIX_Comm_agree and MPIX_Comm_shrink methods. This fall-back implementation is freely available as a separate open source library.[1] It is loaded via the LD_PRELOAD mechanism to overload some MPI_* functions: At the initialization of MPI (MPI_Init or MPI_Init_thread) MPI_COMM_WORLD is internally duplicated with the PMPI_Comm_dup method. This new communicator is used to propagate MPIX_Comm_revoke messages. Then, a non-blocking receive is created with PMPI_Irecv, waiting for these revoke messages. All waiting methods (MPI_Wait, MPI_Waitany, MPI_Waitall, MPI_Waitsome) are extended to test also for revoke messages using PMPI_Waitany. Analogously, all testing methods are extended. If a revoke message is received, the error code MPIX_ERR_REVOKED is returned. All blocking communications are mapped to their non-blocking counterparts followed by MPI_Wait to make sure that they terminate if the communicator is revoked. The MPIX_Comm_agree method is implemented using PMPI_allreduce with MPI_BAND (binary *and* operation) on the communicator that was created in the initialization. The method MPIX_Comm_shrink extracts the group of processes from the revoked communicator using PMPI_Comm_group, and

---

[1] https://gitlab.dune-project.org/exadune/blackchannel-ulfm, BSD-3 licence.

creates a new communicator using `MPI_Comm_create`. It is not possible to use `MPI_Comm_dup`, as there are probably pending collective communication on the revoked communicator, which hinder the collective process of duplication. This implementation is fully compatible with MPI-3.

2.3 Data-Driven Compression with the SZ Library

Compression is natural to consider in numerical algorithms like iterative solvers, as they are inexact by definition: It introduces a small but typically quantifiable or controllable additional error, and the decompressed data is thus different to the input data. We employ the SZ library [17,33,42] for lossy compression in this paper. It was developed with the aim to provide a compression strategy that can significantly reduce the data size while effectively controlling the data distortion. In its current version it features an adaptive compression framework which selects either an improved Lorenzo prediction method or a linear regression method to dynamically compress data in different regions of the dataset. The selection is realised based on the data features in each block to obtain the best compression quality. Because of its prediction/regression-layer it is best suited for structured, e.g., grid based, and ordered datasets. Furthermore the size of the input data must be big enough to show the full potential of the algorithm: The authors recommend to not apply it to datasets with sizes below ten kilobytes.

When applied to suitable data, experiments confirm that SZ yields better compression ratios while having superior rate distortion compared to other lossy compressors [33].

SZ compresses the data in two steps (see Fig. 1): The first step involves an approximation where a first-phase prediction for every new data point is calculated based on already processed data points (dependent on selected layer-size) and a fitting scheme (prediction or regression). Secondly, a Huffmann-like coding is applied to move the first-phase prediction closer to the real value using a small sized integer p (usually 4 bit) providing a second-phase prediction. If the second-phase prediction is not accurate enough, the original data is stored by applying standard binary compression techniques. Subsequent data points then use these more accurate data point in their predictions.

3   Preparing Iterative Linear Solvers for LFLR

Building upon these preliminaries, we introduce a LFLR concept into the DUNE framework, by means of a flexible interface. This especially refers to easy modification of backup storing techniques (location, compression, ...), backup frequency and data to backup and to restore. We recall that we use the iterative solver modules as demonstrator, and do not consider checkpointing matrix or preconditioner data, and neither grid data. See Subsect. 1.3 for possible remedies.

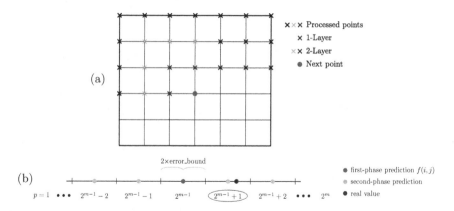

Fig. 1. SZ compression workflow: data is traversed along one grid dimension and values are predicted based on a layer of already processed points (a). Predicted values are shifted by a multiple of a fixed length interval to be as close as possible to the real value (b).

3.1   Backup Creation and Compression

An important aspect of our fault tolerance concept are backups with low memory overhead, which are created locally as far as possible. Thus, backup creation should involve as little communication as possible between participating ranks. Sending local backups to remote ranks introduces a substantial amount of communication, which we alleviate by compression. This communication can additionally be done asynchronously, spread out over multiple solver iterations. We focus on the synchronous case here, but design our interface with asynchronicity in mind.

The interface allows the use of full backups, i.e., copying all desired data as a particular specialization. To reduce the memory and bandwidth footprint of the backups, another specialization of the interface exists for SZ, that can be easily adapted for other compression libraries.

Lossy compression techniques offer better compression rates compared to lossless ones, with the disadvantage of introducing a compression error: For the data x and their decompressed counterpart $\tilde{x}$ it holds that $\tilde{x} = x + eps$ whereby the size of eps can be bounded in some norm depending on the used technique. For iterative solvers which apply convergence control based on the norm of the residual $\|Ax - b\|_2$ with the operator A and the right hand side b, this introduced error is expected to have only negligible influence on the convergence and the solution if it is small enough. This negligible influence even varies during the iterative procedure because the numerical error is decreasing over time and thus the compression error can behave similarly.

Intercepting the iterative solve of the equation $Ax = b$ after iteration i with its current approximation $x^{(i)}$ and replacing it with its decompressed data $\tilde{x}^{(i)}$, we obtain the following in the 2-norm of the residual:

$$\|A\tilde{x}^{(i)} - b\|_2 = \|A(\tilde{x}^{(i)} - x^{(i)} + x^{(i)}) - b\|_2$$
$$= \|Ax^{(i)} + A(\tilde{x}^{(i)} - x^{(i)}) - b\|_2$$
$$\leq \|Ax^{(i)} - b\|_2 + \|A(\tilde{x}^{(i)} - x^{(i)})\|_2 \tag{1}$$

This indicates that the residual norm will maintain its order of accuracy if $\|A(\tilde{x}^{(i)} - x^{(i)})\|_2$ is smaller or of similar magnitude as the fault-free residual norm $\|Ax^{(i)} - b\|_2$. Obviously, as $\|Ax^{(i)} - b\|_2$ decreases towards convergence, $\|A(\tilde{x}^{(i)} - x^{(i)})\|_2$ must also get smaller within the iterative procedure to achieve convergence.

We also investigate different backup frequencies where a backup is only created every n-th iteration and therefore can be outdated in case of local recovery compared to fault-free ranks. This introduces an additional error component into the analysis but enables further opportunities to decrease the computational and numerical overhead of our protection mechanism. This is benefitial in an asynchronous setting, as it allows to distribute the backup creation as well as communication, cf. Subsect. 3.2.

In our implementation, we use the point-wise relative error bound control strategy (PW_REL) provided by the SZ library (cf. Subsect. 2.3). For the current iterate $x^{(i)} \in \mathbb{R}^N$ and the decompressed counterpart $\tilde{x}^{(i)} \in \mathbb{R}^N$ this error control strategy guarantees that for a prescribed tolerance ε it holds that

$$\frac{|\tilde{x}_j^{(i)} - x_j^{(i)}|}{|x_j^{(i)}|} \leq \varepsilon \qquad \forall j = 1, \ldots, N. \tag{2}$$

Summing up the component-wise error over all elements yields after straight-forward calculations:

$$\Rightarrow \qquad \sum_{j=1}^{N} |\tilde{x}_j^{(i)} - x_j^{(i)}|^2 = \|\tilde{x}^{(i)} - x^{(i)}\|_2^2 \leq \varepsilon^2 \|x^{(i)}\|_2^2$$

$$\stackrel{\sqrt{}}{\Longleftrightarrow} \qquad \|\tilde{x}^{(i)} - x^{(i)}\|_2 \leq \varepsilon \|x^{(i)}\|_2$$

This indicates that if Eq. (2) holds true we can guarantee that the global relative error of the decompressed data in the 2-norm is smaller than the prescribed ε.

We can choose ε either fixed over the iterative solve, or couple it adaptively to readily available quantities: Using the operator norm and the induced norm property, we can further estimate the error in Eq. (1) as

$$\|A(\tilde{x}^{(i)} - x^{(i)})\|_2 \leq \|A\|_2 \|\tilde{x}^{(i)} - x^{(i)}\|_2$$

$$= \sup_{x \in \mathbb{R}^N} \frac{\|Ax\|_2}{\|x\|_2} \|\tilde{x}^{(i)} - x^{(i)}\|_2$$

$$\leq \sup_{x \in \mathbb{R}^N} \frac{\|Ax\|_2}{\|x\|_2} \varepsilon \|x^{(i)}\|_2.$$

If we omit the supremum over $\mathbb{R}^N$ and just consider the current $x^{(i)}$ we get

$$\sup_{x \in \mathbb{R}^N} \frac{\|Ax\|_2}{\|x\|_2} \varepsilon \|x^{(i)}\|_2 \approx \frac{\|Ax^{(i)}\|_2}{\|x^{(i)}\|_2} \varepsilon \|x^{(i)}\|_2 = \|Ax^{(i)}\|_2 \varepsilon,$$

while $\|Ax^{(i)}\|_2$ can be bounded by readily available quantities:

$$\|Ax^{(i)}\|_2 = \|Ax^{(i)} - b + b\|_2 \leq \|Ax^{(i)} - b\|_2 + \|b\|_2$$

This means if we couple ε to the current residual 2-norm $\|Ax^{(i)} - b\|_2$ by choosing

$$\varepsilon = \frac{\|Ax^{(i)} - b\|_2}{\|b\|_2 + \|Ax^{(i)} - b\|_2} \mathtt{tol}_{\mathtt{aSZ}}$$

both summands in Eq. (1) are likely to be of similar order of magnitude. Here $\mathtt{tol}_{\mathtt{aSZ}}$ is an additional tuning parameter which can be used to create more or less accurate backups. A lower value of $\mathtt{tol}_{\mathtt{aSZ}}$ will decrease $\|A(\tilde{x}^{(i)} - x^{(i)})\|_2$ but also lower the compression error $\|\tilde{x}^{(i)} - x^{(i)}\|_2$ and thus lead to a lower compression rate and probably an increased compression time. On the other side the influence of the compression error overall will diminish.

This is correlated to a special characteristic of the SZ compression algorithm: If data are not suitable for compression, the compression itself takes longer and yields worse compression rates as when applied to smooth suitable data. In an ideal scenario each value, besides the first ones which are necessary to compute predictions with the chosen prediction scheme, can be represented by a small sized integer. We investigate this particularity of SZ further below.

3.2   Remote Storing of Backups

Due to the high compression rates, we can store backups within the memory of other ranks with an acceptable footprint size. Each rank communicates its local backup to a small subset of neighboring ranks where the backups are kept in memory. The size of this subset is correlated to the requested robustness of the iterative solver. If multiple node losses are expected to happen simultaneously, higher redundancy is necessary. Within our numerical examples we simplify this and just store the backups in a ring topology without duplication because our main research objective so far is to show the general usability of our fault tolerant solver concept.

3.3   Recovery Strategies

Generally, our recovery strategy involves a solver restart combined with the creation of a good initial guess based on non-lost data on fault-free ranks as well as backups on ranks where data was lost. Operations like assembly before the iterative solver have to be recovered in a different way, see Subsects. 1.2 and 1.3.

Based on this assumption we essentially provide three recovery strategies utilising the available backup data: A local rollback, a global rollback and an improved local recovery:

1. **local rollback:** The data on the failed and recovered rank are initialised by their backup data.
2. **global (synchronous) rollback:** The data on all ranks are initialised by backup data corresponding to the same global iteration. Therefore it may be necessary to store multiple backups from each rank and also copies of the rank specific data in case of high fault probability.
3. **improved rollback:** Local data on the recovered rank are initialised by their backup data and post-processed by applying a local auxiliary Dirichlet problem. Here the Dirichlet data are obtained from neighbouring nodes as suggested in our previous publication [25] and Huber et al. [29].

To enable these recovery techniques we also have to store the iteration number (synchronous rollback) as well as the current local residual norm (improved recovery) for each local backup which are negligible scalars.

After the recovery, the new residual norm $\|\tilde{r}^{(i)}\|$ can be split into two parts: Given the set of all participating ranks $\mathcal{R}$ as well as the set $\mathcal{R}_f$ containing all faulty ranks, and indicating the local representation of a vector by a subscript, we obtain the following bound:

$$
\begin{aligned}
\|\tilde{r}^{(i)}\|_2^2 &= \sum_{r\in\mathcal{R}} \|\tilde{r}_r^{(i)}\|_2^2 \\
&\leq \sum_{r\in\mathcal{R}_f} \|A\tilde{x}_r^{(i-1)} - b_r\|_2^2 + \sum_{r\in\mathcal{R}\backslash\mathcal{R}_f} \|Ax_r^{(i)} - b_r\|_2^2 \\
&\leq \sum_{r\in\mathcal{R}_f} \left(\|Ax_r^{(i-1)} - b_r\|_2^2 + \|A(\tilde{x}_r^{(i-1)} - x_r^{(i-1)})\|_2^2 \right) + \sum_{r\in\mathcal{R}\backslash\mathcal{R}_f} \|Ax_r^{(i)} - b_r\|_2^2 \\
&= \sum_{r\in\mathcal{R}_f} \left(\|r_r^{(i-1)}\|_2^2 + \|A(\tilde{x}_r^{(i-1)} - x_r^{(i-1)})\|_2^2 \right) + \sum_{r\in\mathcal{R}\backslash\mathcal{R}_f} \|r_r^{(i)}\|_2^2 \qquad (3)
\end{aligned}
$$

As a consequence, based on the size of $\mathcal{R}$ resp. $\mathcal{R}_f$ and the quality of recovery, the introduced error may be not visible in the residual norm at all. But it still can have an effect on the following iterative convergence procedure and the obtained solution. $\mathcal{R}_f = \mathcal{R}$ corresponds to a global rollback. Applying the improved recovery reduces the target quantity

$$
\sum_{r\in\mathcal{R}_f} \|A(\tilde{x}_r^{(i-1)} - x_r^{(i-1)})\|_2^2
$$

and by this the compression error in Eq. (3).

By design our fault tolerance interface allows to use the created backups to switch to a different type of solver if the previously selected does not yield adequate results.

3.4   Recovery of Iterative Solvers

We want to establish recovery approaches which store as little data as possible while still providing better performance than simple restart mechanisms. The simplest approach which can be applied to any iterative solver is a restart with an initial guess which uses the progress which has been made until the local rank loss. Ranks which did not die have their data present while for the crashed nodes backups are remotely available. Utilizing these data it is possible to build up a new global initial guess for any solver.

In the spirit of the ABFT concept, there is a lot of headspace for improvement:

CG with Rollback: The Conjugate Gradients method [27] is a widely used algorithm to solve problems arising from linear partial differential equations with sparse symmetric positive definite operators. A great advantage of the CG method is that it makes use of a short recurrence. For CG a restart involves rebuilding its associated Krylov space to which the iterate $x^{(i)}$ was orthogonalized before. When restarting the solver with the recovered iterate $\tilde{x}^{(i)}$ all this information is lost. This means, that in case of a rollback the solver builds a new Krylov space

$$\mathcal{K}^k(A, \tilde{r}^{(i)}) = \mathrm{span}(\tilde{r}^{(i)}, ..., A^{k-1}\tilde{r}^{(i)}), \tag{4}$$

where $\tilde{r}^{(i)} = b - A\tilde{x}^{(i)}$ is the residual of the recovered solution. Note that this is implicit due to the short recurrence. In case of our rollback recoveries, the initial guess is either recovered locally from its backup (local rollback) or globally (global rollback). In case of the global rollback the local parts of the global iterate are from the same iteration while a local rollback combines data from different previous iterations.

Full CG Recovery: In order to avoid this restart and keep information of previous iterations we can make use of the short recurrence in our recovery strategy. We not only checkpoint the current iterate but also the corresponding search direction which is intended for the following iteration.

In case of a recovery, this search direction is then used to build up the implicit Krylov space. All subsequent search directions are computed from the CG relation

$$p^{(k)} = r^{(k)} - \beta p^{(k-1)} \qquad\qquad \forall k = i+1, \dots. \tag{5}$$

In case of an exact backup, i.e. $\tilde{p}^{(i)} = p^{(i)}$, this space is orthogonal to the previous Krylov space

$$\mathcal{K}^{i-1}(A, r^{(0)}) = \mathrm{span}\left(p^{(0)}, \dots, p^{(i-1)}\right). \tag{6}$$

Therefore, the iteration sequence is arithmetically the same as in the fault-free case.

Storing the search direction next to the iterate proves to be particularly advantageous in the case of multiple data losses, as seen in our numerical tests. A requirement to make use of this advantage is a global synchronous rollback. Otherwise the local iterates and search directions may belong to different iterations and the solver behaves more or less similar to the scenario where no information besides the iterate are kept.

Our numerical tests show, that even with a moderate compression accuracy, the iteration count is only mildly affected.

GMRES Recovery: In contrast to the Conjugate Gradients method GMRES [37] is never applied in its full version for relevant applications, due to prohibitive memory requirements resulting from the lack of a short-term recurrence: For each iteration one additional full basis vector has to be stored. Furthermore the solution is computed by solving a minimization problem combining all these basis vector at the end of the overall solve. Instead, the so-called GMRES(m) version is used where the GMRES algorithm is terminated after m iterations, followed by a computation of an approximate solution and a restart of the same algorithm with it as initial guess.

Because a restart is already involved, we simply restart GMRES(m) in case of a node loss, utilizing the backup data. This means that we lose, in the worst-case, one m-block of the GMRES(m) algorithm because no more information is shared within different blocks beside the iterate. We expect that an additional local compression error only shows minimal effect for this recovery approach.

In theory it would also be possible to store the basis but the storage costs are quite significant and hard to justify by the only moderate advantages.

4   Implementation for Iterative Solvers in DUNE

The iterative solvers in DUNE are developed with dynamic polymorphism to make them configurable during runtime and make it possible to formulate the algorithms in an abstract framework of operator/preconditioner applications and scalar product computations. Every `IterativeSolver` holds `std::shared_ptr` to the abstract interfaces `LinearOperator`, `Preconditioner` and `ScalarProduct`. These interfaces can be implemented for different use cases and parallelization strategies.

All code is available from the project gitlab in the specific feature branches.[2]

4.1   Framework Extensions

We extend this framework by an abstract class `BackupRestoreManager`. With this class the solver is able to register backups during the solution process. Furthermore, the solver can check during the setup phase, whether a backup from a previous run is available. Depending on the backup strategy this interface can be implemented in different ways: For our tests, we have implemented the `RemoteBackupRestoreManager`, which distributes the in-memory backups in remote processes using a ring topology, i.e., backups of rank r are stored on rank $(r+1)$ mod s, where s is the number of ranks. Losing two or more consecutive ranks yields at least one rank without available backup, and a possible remedy is to store backups with more redundancy, or to fall back to synchronous global checkpoints on disk, depending on the application.

---

[2] https://gitlab.dune-project.org/exadune/dune-common/tree/feature/ulfm-mpigu ard, https://gitlab.dune-project.org/exadune/dune-istl/tree/fault_tolerance_interf ace.

To make it possible to switch between multiple compression techniques, we introduce a further abstract interface `Compressor`. Classes that satisfy the `Compressor` interface can be used to compress and restore multiple vectors and auxiliary data in/from a backup. The compression accuracy can be specified by passing a double. The signatures of the most important methods are:

```
1  void compress(const X& vector, double accuracy=1.);
2  void compress(unsigned char* data, size_t len, double eps=1.);
3  void restore(X& vector);
4  void restore(unsigned char* data, size_t len);
```

Furthermore, the class provides information whether the compression is lossy and the compressed data is of constant size.

For our tests we implemented a `CopyCompressor`, that copies the data in memory, and a `SZCompressor`, that uses the SZ library.

4.2 Modifications for CG and GMRES(m)

To integrate the classes into the CG solver, we need to adapt the algorithm at two places. In the initialization phase, we check whether a backup is available and if so we restore it. During the restore we distinguish between a simple and full backup. The `CGSolver` stores internally a `std::shared_ptr<BackupRestoreManager<Vector>>` called `_brm`; in the code snippets below to access the backup. The following listing show a simplified code.

```
1  void CGSolver::apply(Vector& x, Vector& b){
2    // prepare solver
3    int i=1;
4    ...
5    // try recovery
6    if(_brm->isRestart()){
7      bool onlyXrestored = _brm->restore(i, def, {x, p});
8      _prec->pre(x,b);                // makes x consistent
9      i++;
10     _op->applyscaleadd(-1,x,b);    // overwrite b with residual
11     if(onlyXrestored){
12       p=0;
13       _prec->apply(p,b);            // apply preconditioner
14     }
15   }else{
16     // compute initial residual
17     ...
18   }
19   ...
20   // the actual loop
21   for ( ; i<=_maxit; i++ ) {
22     ...
23   }
24 }
```

To create the backup we introduce the following call after the new search direction p is computed.

```
1  backup_future = _brm->createBackup(i,def,{x,p});
```

This creates a backup of the current iteration number i, the current defect def and the vectors x and p. Note that p is only stored if the environment is configured accordingly. The returned backup_future is an object, which can encapsulate communication, as relevant for the asynchronous case.

5   Numerical Examples

In our experiments, we first examine the numerical behavior of the different strategies introduced above. Then, we quantify the overhead introduced by the compression and additional communication. Small-scale test problems with single and multiple injected faults are run for viability analysis, and large-scale fault-free tests for performance analysis. The former are designed as a worst case, with a fault injection scheme that is unrealistic in future systems, but which demonstrates the limitations in terms of applicability.

5.1   Small-Scale Viability Tests

The small-scale test problem is a Poisson problem with the known solution $u(x,y) = e^{-x^2-y^2}$ and corresponding right hand side b:

$$-\Delta u = b \qquad\qquad \text{in } \Omega = (0,1)^2$$
$$u(x,y) = e^{-x^2-y^2} \qquad\qquad \text{on } \partial\Omega$$

It is discretized with bilinear Lagrange Finite Elements on a 300×300 square grid, partitioned with minimal overlap to four ranks. For solving we use the CG resp. GMRES(20) method with an ILU(1) preconditioner until the initial residual is reduced by a factor of 10^{-6}. When faults are injected multiple times the rank which loses the data is traversed in a cyclic fashion.

Conjugate Gradients: Solving this problem with the CG method needs 251 iterations for reaching the prescribed reduction in the fault-free scenario. Table 1 shows the number of iterations which are needed when different recovery approaches are combined, using either a Copy backup or SZ backups with different accuracies tol_{aSZ}, both only for the iterate $x^{(i)}$. A fault, i.e., data loss, is injected in iteration 60, and then in every 20th iteration: We thus consider the unrealistically hard case of very frequent failures.

As expected we see that both local and global rollbacks lead to a significantly increased iteration count because the solver is restarted every 20th iteration and loses all information about its Krylov space history. Due to this information loss, the amount of iterations needed for convergence is increased by a factor of four (1022 iterations) if the solver is rolled back synchronously and no compression is involved. Introducing an additional compression error does not lead to a noteworthy additional increase.

Table 1. Iteration count for different recovery strategies and compression accuracies, conjugate gradients. n/c denotes non-converging schemes.

Strategy	Copy	$\texttt{tol}_{\text{aSZ}}$				
		1	10^{-1}	10^{-2}	10^{-3}	10^{-4}
Local rollback	890	951	931	888	890	890
Global rollback	1022	1029	1024	1022	1022	1022
Full recovery	251	n/c	n/c	283	253	253

Compared to the global rollback, a local rollback results in a smaller increase of iterations numbers as it keeps the progress which has been made on the fault-free ranks. In combination with a compression error these numbers can increase but selecting a $\texttt{tol}_{\text{aSZ}}$ of 10^{-2} and below reduces this compression error to have a diminishing effect. Choosing such a tolerance still gives compression rates between 20–30 in our experiments. Since SZ compression is designed for high resolution data sets and because in our case the local test problems have only 22500 degrees of freedom it is expected to give far better compression rates when used beyond toy problems.

Because of the high fault rate and the subsequent information loss, applying the auxiliary problem, i.e., the improved recovery in CG, is not improving the results significantly, and is therefore omitted in the table. It can only be used to increase the compression rate further by reducing $\texttt{tol}_{\text{aSZ}}$. In such a case the improved recovery gives similar results to Copy backups but with more recovery overhead.

In particular, the 'local' scheme shows that compressing the iterate $x^{(i)}$ does not deteriorate convergence too much, compared with copy compression. The results obtained for the global scheme(s) show that not changing the backup locally but rerolling is worse, because less information on the iterate is kept, irrespective of compression accuracy.

The scheme specifically tailored for CG (labeled 'full') indicates that a too strong compression (values of $\texttt{tol}_{\text{aSZ}} \geq 10^{-2}$) of not only the iterate $x^{(i)}$, but also the search direction $p^{(i)}$ leads to a situation where restarting the implicit building of the Krylov space means that the two no longer seem to suffice to build the global approximation space, as the compression is oblivious of the different roles of the iterate and the search direction in CG.

Overall we observe that for the CG method storing the search direction for recovery can improve the solver robustness significantly when multiple data losses are expected because it is possible to keep the information from the previous iterations. Using a moderate accuracy ($\texttt{tol}_{\text{aSZ}}$) even makes compression viable. On the other hand storing the search direction introduces additional bandwidth- and memory-overhead, which is however alleviated by compression.

GMRES(m): As explained in Sect. 3.4, GMRES(m) has inherent restarts and does not have a true iterate update within each block of m iterations. We consider

Table 2. Iteration count for different recovery strategies and compression accuracies, GMRES(20).

Strategy	Copy	tol_{aSZ}				
		1	10^{-1}	10^{-2}	10^{-3}	10^{-4}
Local rollback	555	561	557	557	556	555
Global rollback	555	583	577	553	557	555

two recovery approaches for this algorithm: Local and global rollback. Each of them restarts the GMRES(m) solver with the last available data and loses all progress made within one m-block. This simulates a fault injection at any position in the faulty block.

Table 2 shows the iteration numbers for a test scenario in which the fault-free run takes 555 iteration until convergence. This means that the algorithm terminates after the 28th m-block. Here faults are injected every 4th block, i.e. after 60 successful iterations, beginning at the end of the fifth m-block.

Because GMRES never updates the iterate within one m-block, a fault within any iteration in this block yields the same result after recovery: The solver is restarted with the approximation computed by the last successful m-block which is still locally available or can, for the faulty ranks, be received from a remote rank involving a compression error depending on the compression accuracy. Thus, the only difference between a local rollback and a global rollback is that the compression error is introduced globally in case of the second approach. Furthermore it should be obvious that a global rollback would never be applied in practice when using compression because the fault-free ranks still have access to their uncompressed data. It only demonstrates the worst case where a lot of ranks lose their data.

Taking a closer look at the results we see that the introduced local compression error does not have a big effect on the necessary iteration number to reach the convergence criterion. Even selecting $tol_{aSZ} = 10^0$ provides suitable backup data for recovery while yielding compression rates of 40 and above. In addition it illustrates the robustness of GMRES(m) as introducing even global compression errors after every 4th m-block do not deteriorate convergence further.

For GMRES(m) we can conclude that due to its properties, i.e., higher robustness and inherent restarts, it seems to be more suitable for our local recovery approaches than the Conjugate Gradients method. Although we can not store any information from an m-block, the number of successful iterations needed for convergence does not change in comparison to the fault-free scenario.

Conjugate Gradients with Occasional Faults and Older Backups: The previous scenario has by far higher fault rates than expected for future systems. As seen in Table 3 the local recovery approach is more competitive if data loss occurs only occasionally, even when the backup frequency is low and outdated backups are used. We introduce a single data loss at iterations $i \in \{20, 60, 120, 200\}$, and the recovery is initiated with a backup from the last iterate (left) resp. a ten iterations old backup (right). Note that even full recovery with a Copy backup of the age ten increases the iteration count by nine since progress of iterations between backup and data-loss is lost. This means that, when a ten iteration old Copy backup is used and data is lost after iteration $i = 60$, the solver effectively restarts after iteration 50.

Table 3. Iteration numbers for convergence of CG with single data-losses and recovery based on a one (left) resp. ten (right) iterations old backup.

			Iteration							Iteration			
			20	60	120	200				20	60	120	200
Full recovery	tol_{aSZ}	Copy	251	251	251	251	Full recovery	tol_{aSZ}	Copy	260	260	260	260
		1	253	281	655	1556			1	262	291	381	1592
		10^{-1}	253	281	341	272			10^{-1}	260	291	369	284
		10^{-2}	253	281	253	253			10^{-2}	262	263	265	262
Global rollback	tol_{aSZ}	Copy	251	281	299	263	Global rollback	tol_{aSZ}	Copy	260	270	306	265
		1	251	281	299	265			1	260	270	307	267
		10^{-1}	251	281	299	264			10^{-1}	260	270	307	266
		10^{-2}	251	281	299	263			10^{-2}	260	270	306	265
Local rollback	tol_{aSZ}	Copy	252	282	296	264	Local rollback	tol_{aSZ}	Copy	261	288	287	268
		1	252	282	298	264			1	261	288	287	269
		10^{-1}	252	282	298	264			10^{-1}	261	288	287	268
		10^{-2}	252	282	296	264			10^{-2}	261	287	287	268
Improved	tol_{aSZ}	Copy	249	278	298	263	Improved	tol_{aSZ}	Copy	258	267	289	265
		1	249	278	298	263			1	257	267	289	265
		10^{-1}	249	278	298	263			10^{-1}	257	267	289	265
		10^{-2}	249	278	298	263			10^{-2}	257	267	289	265

In both scenarios, the full recovery without compression (Copy) offers the best results with respect to iteration numbers because a local recovery still always loses the Krylov space information. Setting the compression target below 10^{-2} does not change the results, so we omit these tests in the table. However in contrast to the previous fault scenario all rollback recoveries increase the iteration number only marginally even when less accurate compression is involved. Surprisingly sometimes an older backup can be superior to a recent one (see data loss at iteration 60 with global rollback or data-loss at iteration 120 with local rollback), and we assume numerical noise and granularity effects. Applying an improved recovery with ten local CG iterations without a preconditioner for the auxiliary problem reduces the amount of necessary global iterations further, especially when an older backup is used.

5.2   Overhead Quantification at Scale

To quantify the overhead of the backup strategy, we run test simulations on the PALMAII cluster of the University of Münster. We do not inject artificial errors, i.e., no restore is performed. The test is performed in weak scaling, with approx. 10^6 DoF per rank. Backups of the current solution approximation $x^{(i)}$ are created at every CG iteration. The overhead, comprising both the copy or compression to create the backup as well as sending it to the next rank, is shown in Fig. 2. The backup is sent in the same cyclic fashion to a successor as before. On a single rank we do not have any communication overhead but only the necessity to create a local copy of the data. We observe that these copy operation yields an overhead of less than 2% when no compression is applied. Utilizing compression increases the overhead to 10–12% because of the additional time needed to compress the data. When more and more ranks, i.e. communication, are involved the overhead of the compression goes down significantly and it nearly reaches the 'no compression' case. Considering scenarios with an overall higher bandwidth pressure and/or increased backup redundancy resulting in more communication, we expect the utilization of compression techniques to become superior. In addition, the compression overhead of SZ reduces further for smoother and/or higher resolved data. In our experiments, the interconnect of the cluster is 100 Gbit/s Intel Omni-Path, which is not the limiting resource. In addition, when backups are communicated to further ranks for higher robustness the compression overhead will diminish because the communication overhead for the compressed backups will increase to a lesser extent than for the uncompressed backup.

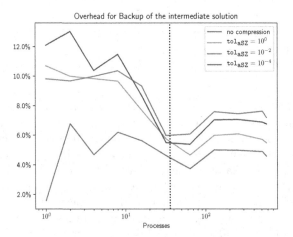

Fig. 2. Overhead of backup strategies. Backup of intermediate solution in every iteration, with different compression methods. The vertical dotted line marks the process number of one node.

6 Conclusions

We have designed an abstract interface that enables the integration of LFLR techniques into linear solvers. The implementation in DUNE is flexible, and has been specialized so far for CG and GMRES(m). In the spirit of ABFT, there exists both basic functionality (as sufficient for GMRES(m)), and we exploit solver-specific properties to improve the schemes.

We have evaluated our approach and demonstrated sufficiently small overhead. In terms of numerical performance, local recovery seems to be promising when faults do not appear too frequently, which is a realistic assumption for future machines. For higher fault rates GMRES is more robust to errors and local recovery than CG. CG recovery can be improved by storing its search direction and thus keeping the Krylov space information.

First observations show that with single backup propagation (i.e., neglecting redundancy of the backups), SZ compression introduces a marginal overhead compared to a full checkpoint propagation, but essentially enables in-memory backups on neighbouring nodes.

The next step is to extend our LFLR techniques to other modules in DUNE.

Acknowledgements. Supported by the German Research Foundation in the Priority Programme 1648 'Software for Exascale Computing', grants GO 1758/2-2 and EN 1042/2-2; and under Germany's Excellence Strategy EXC 2044–390685587, Mathematics Münster: Dynamics–Geometry–Structure.

References

1. Agullo, E., Giraud, L., Guermouche, A., Roman, J., Zounon, M.: Numerical recovery strategies for parallel resilient Krylov linear solvers. Numer. Linear Algebra Appl. **23**(5), 888–905 (2016)
2. Agullo, E., Giraud, L., Salas, P., Zounon, M.: Interpolation-restart strategies for resilient eigensolvers. SIAM J. Sci. Comput. **38**(5), C560–C583 (2016)
3. Ali, M.M., Southern, J., Strazdins, P., Harding, B.: Application level fault recovery: using Fault-Tolerant Open MPI in a PDE solver. In: 2014 IEEE International Parallel and Distributed Processing Symposium Workshops (IPDPSW), pp. 1169–1178. IEEE (2014)
4. Altenbernd, M., Göddeke, D.: Soft fault detection and correction for multigrid. Int. J. High Perform. Comput. Appl. **32**(6), 897–912 (2018). https://doi.org/10.1177/1094342016684006
5. Ashraf, R.A., Hukerikar, S., Engelmann, C.: Shrink or substitute: handling process failures in HPC systems using in-situ recovery. In: Proceedings of the 26th Euromicro International Conference on Parallel, Distributed and Network-based Processing (PDP 2018), pp. 178–185. IEEE (2018)
6. Bastian, P., et al.: A generic grid interface for parallel and adaptive scientific computing. Part II: implementation and tests in DUNE. Computing **82**(2–3), 121–138 (2008). https://doi.org/10.1007/s00607-008-0004-9
7. Bastian, P., et al.: A generic grid interface for parallel and adaptive scientific computing. Part I: abstract framework. Computing **82**(2–3), 103–119 (2008). https://doi.org/10.1007/s00607-008-0003-x

8. Bastian, P., et al.: The DUNE framework: basic concepts and recent developments. arXiv preprint arXiv:1909.13672 (2019)
9. Bautista-Gomez, L., Zyulkyarov, F., Unsal, O., McIntosh-Smith, S.: Unprotected computing: a large-scale study of dram raw error rate on a supercomputer. In: Proceedings of the International Conference for High Performance Computing, Networking, Storage and Analysis, p. 55. IEEE Press (2016)
10. Bland, W., Lu, H., Seo, S., Balaji, P.: Lessons learned implementing user-level failure mitigation in MPICH. In: Proceedings of the 2015 15th IEEE/ACM International Symposium on Cluster, Cloud and Grid Computing, pp. 1123–1126 (2015). https://doi.org/10.1109/CCGrid.2015.51
11. Bland, W., Bouteiller, A., Herault, T., Hursey, J., Bosilca, G., Dongarra, J.J.: An evaluation of user-level failure mitigation support in MPI. In: Träff, J.L., Benkner, S., Dongarra, J.J. (eds.) EuroMPI 2012. LNCS, vol. 7490, pp. 193–203. Springer, Heidelberg (2012). https://doi.org/10.1007/978-3-642-33518-1_24
12. Blatt, M., Bastian, P.: The iterative solver template library. In: Kågström, B., Elmroth, E., Dongarra, J., Waśniewski, J. (eds.) PARA 2006. LNCS, vol. 4699, pp. 666–675. Springer, Heidelberg (2007). https://doi.org/10.1007/978-3-540-75755-9_82
13. Cantwell, C.D., Nielsen, A.S.: A minimally intrusive low-memory approach to resilience for existing transient solvers. J. Sci. Comput. $78(1)$, 565–581 (2019)
14. Cappello, F.: Fault tolerance in petascale/exascale systems: current knowledge, challenges and research opportunities. Int. J. High Perform. Comput. Appl. $23(3)$, 212–226 (2009)
15. Cappello, F., Geist, A., Gropp, W., Kale, S., Kramer, B., Snir, M.: Toward exascale resilience: 2014 update. Supercomputing Front. Innovations $1(1)$, 5–28 (2014)
16. Chen, C., Du, Y., Zuo, K., Fang, J., Yang, C.: Toward fault-tolerant hybrid programming over large-scale heterogeneous clusters via checkpointing/restart optimization. J. Supercomputing $75(8)$, 4226–4247 (2017). https://doi.org/10.1007/s11227-017-2116-5
17. Di, S., Cappello, F.: Fast error-bounded lossy HPC data compression with SZ. In: Proceedings of the 2016 IEEE International Parallel and Distributed Processing Symposium, pp. 730–739. IEEE (2016)
18. Di Martino, C., Kramer, W., Kalbarczyk, Z., Iyer, R.: Measuring and understanding extreme-scale application resilience: a field study of 5,000,000 HPC application runs. In: Proceedings of the 2015 45th Annual IEEE/IFIP International Conference on Dependable Systems and Networks, pp. 25–36. IEEE (2015)
19. Dongarra, J., Herault, T., Robert, Y.: Fault tolerance techniques for high-performance computing. In: Herault, T., Robert, Y. (eds.) Fault-Tolerance Techniques for High-Performance Computing. CCN, pp. 3–85. Springer, Cham (2015). https://doi.org/10.1007/978-3-319-20943-2_1
20. Dongarra, J., et al.: The international exascale software project roadmap. Int. J. High Perform. Comput. Appl. $25(1)$, 3–60 (2011). https://doi.org/10.1177/1094342010391989
21. Dongarra, J., et al.: Applied mathematics research for exascale computing. Technical report, U.S. Department of Energy, Office of Science, Advanced Scientific Computing Research Program (2014). http://science.energy.gov/~/media/ascr/pdf/research/am/docs/EMWGreport.pdf
22. Engwer, C., Altenbernd, M., Dreier, N.A., Göddeke, D.: A high-level C++ approach to manage local errors, asynchrony and faults in an MPI application. In: Proceedings of the 26th Euromicro International Conference on Parallel, Distributed and Network-Based Processing (PDP 2018), pp. 714–721. IEEE (2018)

23. Gamell, M., et al.: Evaluating online global recovery with fenix using application-aware in-memory checkpointing techniques. In: Proceedings of the 45th International Conference on Parallel Processing Workshops (ICPPW 2016), pp. 346–355. IEEE (2016)

24. Gamell, M., Katz, D., Kolla, H., Chen, J., Klasky, S., Parashar, M.: Exploring automatic, online failure recovery for scientific applications at extreme scales. In: Proceedings of the International Conference for High Performance Computing, Networking, Storage and Analysis, SC 2014, pp. 895–906. IEEE (2014)

25. Göddeke, D., Altenbernd, M., Ribbrock, D.: Fault-tolerant finite-element multigrid algorithms with hierarchically compressed asynchronous checkpointing. Parallel Comput. **49**, 117–135 (2015)

26. Gupta, S., Patel, T., Engelmann, C., Tiwari, D.: Failures in large scale systems: long-term measurement, analysis, and implications. In: Proceedings of the International Conference for High Performance Computing, Networking, Storage and Analysis, p. 44. ACM (2017)

27. Hestenes, M.R., Stiefel, E.: Methods of conjugate gradients for solving linear systems. J. Res. Natl. Bur. Stand. **49**(6), 409–436 (1952)

28. Huang, K.H., Abraham, J.: Algorithm-based fault tolerance for matrix operations. IEEE Trans. Comput. **100**(6), 518–528 (1984)

29. Huber, M., Gmeiner, B., Rüde, U., Wohlmuth, B.: Resilience for massively parallel multigrid solvers. SIAM J. Sci. Comput. **38**(5), S217–S239 (2016)

30. Keyes, D.E.: Exaflop/s: the why and the how. Comptes Rendus Mécanique **339**(2–3), 70–77 (2011). https://doi.org/10.1016/j.crme.2010.11.002

31. Kohl, N., et al.: A scalable and extensible checkpointing scheme for massively parallel simulations. Int. J. High Perform. Comput. Appl. **33**, 571–589 (2017). https://doi.org/10.1177/1094342018767736

32. Langou, J., Chen, Z., Bosilca, G., Dongarra, J.: Recovery patterns for iterative methods in a parallel unstable environment. SIAM J. Sci. Comput. **30**, 102–116 (2007)

33. Liang, X., et al.: Error-controlled lossy compression optimized for high compression ratios of scientific datasets. In: Proceedings of the IEEE International Conference on Big Data (Big Data 2018), pp. 438–447 (2018)

34. Losada, N., Bosilca, G., Bouteiller, A., González, P., Martín, M.: Local rollback for resilient MPI applications with application-level checkpointing and message logging. Future Gener. Comput. Syst. **91**, 450–464 (2019)

35. Meuer, H., Strohmaier, E., Dongarra, J.J., Simon, H.D.: Top500 supercomputer sites (2019). http://www.top500.org/

36. Nielsen, A.S.: Scaling and resilience in numerical algorithms for exascale computing. Ph.D. thesis, École Polytechnique Fédérale de Lausanne (2018). https://infoscience.epfl.ch/record/258087/files/EPFL_TH8926.pdf

37. Saad, Y., Schultz, M.H.: GMRES: a generalized minimal residual algorithm for solving nonsymmetric linear systems. SIAM J. Sci. Stat. Comput. **7**(3), 856–869 (1986). https://doi.org/10.1137/0907058

38. Schroeder, B., Gibson, G.: A large-scale study of failures in high-performance computing systems. IEEE Trans. Dependable Secure Comput. **7**(4), 337–350 (2009)

39. Sloan, J., Kumar, R., Bronevetsky, G.: An algorithmic approach to error localization and partial recomputation for low-overhead fault tolerance. In: Dependable Systems and Networks (DSN 2013), pp. 1–12 (2013). https://doi.org/10.1109/DSN.2013.6575309

40. Snir, M., et al.: Addressing failures in exascale computing. Int. J. High Perform. Comput. Appl. **28**(2), 129–173 (2014)

41. Snir, M., et al.: Addressing failures in exascale computing. Int. J. High Perform. Comput. Appl. **28**(2), 129–173 (2014). https://doi.org/10.1177/1094342014522573
42. Tao, D., Di, S., Chen, Z., Cappello, F.: Significantly improving lossy compression for scientific data sets based on multidimensional prediction and error-controlled quantization. In: Proceedings of the IEEE International Parallel and Distributed Processing Symposium (IPDPS 2017), pp. 1129–1139. IEEE (2017)
43. Teranishi, K., Heroux, M.A.: Toward local failure local recovery resilience model using MPI-ULFM. In: Proceedings of the 21st European MPI Users' Group Meeting, p. 51. ACM (2014)

Complexity Analysis of a Fast Directional Matrix-Vector Multiplication

Günther Of[✉][iD] and Raphael Watschinger[iD]

Institute of Applied Mathematics, Graz University of Technology,
Steyrergasse 30, 8010 Graz, Austria
of@tugraz.at, watschinger@math.tugraz.at

Abstract. We consider a fast, data-sparse directional method to realize matrix-vector products related to point evaluations of the Helmholtz kernel. The method is based on a hierarchical partitioning of the point sets and the matrix. The considered directional multi-level approximation of the Helmholtz kernel can be applied even on high-frequency levels efficiently. We provide a detailed analysis of the almost linear asymptotic complexity of the presented method. Our numerical experiments are in good agreement with the provided theory.

Keywords: Helmholtz · Fast multipole method · Hierarchical matrix

1 Introduction

In this paper we consider an efficient method for the computation of the matrix-vector product for a fully populated matrix $A \in \mathbb{C}^{N_T \times N_S}$ with entries

$$A[j,k] = f(x_j, y_k), \tag{1}$$
$$f(x,y) = \frac{\exp(i\kappa|x-y|)}{4\pi|x-y|}$$

where f is the Helmholtz kernel, $\kappa > 0$ the wave number and $P_T = \{x_j\}_{j=1}^{N_T}$ and $P_S = \{y_k\}_{k=1}^{N_S}$ are two sets of points in $\mathbb{R}^3$. Similar matrices arise in the solution of boundary value problems for the Helmholtz equation by boundary element methods. Using standard matrix-vector multiplication is prohibitive for large N_T and N_S due to the asymptotic runtime and storage complexity $\mathcal{O}(N_T N_S)$.

Due to the oscillating behavior of the Helmholtz kernel, existing standard fast methods for the reduction of the complexity do not perform well for relatively large wave numbers κ. Therefore, a variety of methods have been developed. There are several versions of the fast multipole method (FMM) based on different expansions of the Helmholtz kernel f. A first version suitable for high frequency regimes is given in [16] and an overview of the early developments can be found in [14]. Of further interest are the methods in [8,12], which rely on plane wave expansions, and the wideband method in [7] which switches between different expansions in low and high frequency regimes.

T. Kozubek et al. (Eds.): HPCSE 2019, LNCS 12456, pp. 39–59, 2021.
https://doi.org/10.1007/978-3-030-67077-1_3

Directional methods allow to overcome the deficiencies of standard schemes in high frequency regimes, too. The basic idea of these methods is that the Helmholtz kernel f can locally be smoothed by a plane wave. In the context of fast methods this idea was first considered in [6] and later in [9]. In [13] the idea is picked up and combined with an approximation of the kernel via interpolation. [3,5] follow a similar path in the context of $\mathcal{H}^2$-matrices providing a rigorous analysis. A slightly different method is proposed in [1], where the directional smoothing is combined with a nested cross approximation of the kernel.

In this paper we present a directional method in the spirit of [13] based on a uniform clustering of the point sets. We choose this approach due to the applicability of the involved interpolation to other kernels and a smooth transition between low and high frequency regimes in contrast to the wideband FMM in [7]. We give a description of the method in Sect. 2 and an asymptotic complexity analysis in Sect. 3. While [13] provides already a brief analysis we present a detailed one not unlike the one in [3], but focusing on points distributed in 3D volumes instead of points on 2D manifolds and allowing two distinct sets of points. In addition, we exploit the uniformity for a significant storage reduction compared to non-uniform approaches. This reduction and the claimed almost linear asymptotic behavior can be observed in our numerical tests in Sect. 4.

2   Derivation of the Fast Directional Method

In this section we present a method for fast matrix-vector multiplications for the matrix A in (1) based on a hierarchical partitioning of the sets of points into boxes and a directional multi-level approximation of the Helmholtz kernel f on suitable pairs of such boxes.

2.1   Box Cluster Trees

The desired matrix partition can efficiently be constructed from a hierarchical tree clustering of the point sets into axis-parallel boxes. In what follows we define uniform box cluster trees which are constructed by a uniform subdivision of an initial box, see, e.g., [10]. In particular, we construct a uniform box cluster tree $\mathcal{T}_T$ for a given set of points $P_T = \{x_j\}_{j=1}^{N_T}$ in an axis-parallel box $T = (a_1, b_1] \times \ldots \times (a_3, b_3] \subset \mathbb{R}^3$ by Algorithm 1. As additional parameter we have the maximal number of points per leaf $n_{\max}$. We use standard notions of levels and leaves in trees known from graph theory. In addition we define

- the *index set* $\hat{t} := \{j \in \{1, \ldots, N_T\} \colon x_j \in t\}$ for a box $t \in \mathcal{T}_T$,
- the *level sets of the tree* by $\mathcal{T}_T^{(\ell)} := \{t \in \mathcal{T}_T \colon \text{level}(t) = \ell\}$,
- the *depth* $p(\mathcal{T}_T) := \max\{\text{level}(t) : t \in \mathcal{T}_T\}$ of the cluster tree $\mathcal{T}_T$,
- the set $\mathcal{L}_T$ of all *leaves* of $\mathcal{T}_T$.

In general, Algorithm 1 creates an adaptive, i.e. unbalanced cluster tree depending on the point distribution. Other construction principles for box cluster trees

---

Algorithm 1. Construction of a uniform box cluster tree $\mathcal{T}_T$

1: **input**: Points $P_T = \{x_j\}_{j=1}^{N_T}$ inside a box $T = (a_1, b_1] \times \ldots \times (a_3, b_3]$, maximal number $n_{\max}$ of points per leaf.

2: Construct an empty tree $\mathcal{T}_T$ and add T as its root.

3: Call REFINECLUSTER(T, $\mathcal{T}_T$)

4: **function** REFINECLUSTER($T = (a_1, b_1] \times \ldots \times (a_3, b_3]$, $\mathcal{T}$)

5:      **if** $\#\{x_j : x_j \in T\} > n_{\max}$ **then**

6:          Compute center $c_1 = (a_1 + b_1)/2$, $c_2 = (a_2 + b_2)/2$, $c_3 = (a_3 + b_3)/2$.

7:          Uniformly subdivide T into 8 boxes $T_1 = (a_1, c_1] \times \ldots \times (a_3, c_3], \ldots,$

8:          $T_8 = (c_1, b_1] \times \ldots \times (c_3, b_3]$.

9:          **for** $k = 1, \ldots, 8$ **do**

10:             **if** $\#\{x_j : x_j \in T_k\} \geq 1$ **then**

11:                Add T_k to $\mathcal{T}$ as child of T.

12:                Call REFINECLUSTER(T_k, $\mathcal{T}$).

---

such as bisection [15, Sect. 3.1.1] tailor the tree to the point sets yielding more balanced trees. However, the boxes at a given level ℓ of such a tree can vary strongly in shape, while the ones of a uniform box cluster tree are identical up to translation. We will exploit this uniformity to avoid recomputations and to reduce the storage costs of the presented method.

2.2 A Directional Kernel Approximation

In this section we describe a method to approximate the Helmholtz kernel f on a suitable pair of boxes t and s by a separable expansion, which will allow for low rank approximations of suitable subblocks of the matrix A in (1). Due to the oscillatory part $\exp(i\kappa|x - y|)$ of f, standard approaches like tensor interpolation of the kernel are not effective for relatively large κ as pointed out in [1,13]. Therefore, we consider a directional approach which first appeared in [6] and [9] and was later used in [5] and [13] among others. The basic idea is that the oscillatory part $\exp(i\kappa|x - y|)$ of f can be smoothened by a plane wave term $\exp(-i\kappa\langle x-y, c\rangle)$ in a cone around a direction $c \in \mathbb{R}^3$ with $|c| = 1$. We can rewrite the Helmholtz kernel f by expanding the numerator and the denominator by a plane wave term yielding

$$f(x, y) = f_c(x, y) \exp(i\kappa\langle x, c\rangle) \exp(-i\kappa\langle y, c\rangle), \tag{2}$$

$$f_c(x, y) := f(x, y) \exp(-i\kappa\langle x - y, c\rangle) = \frac{\exp(i\kappa(|x - y| - \langle x - y, c\rangle))}{4\pi|x - y|}. \tag{3}$$

The modified kernel function f_c is somewhat smoother than f on suitable boxes t and s. In fact, if two points $x \in t$ and $y \in s$ satisfy $(x - y)/|x - y| \approx c$, then $f_c(x, y) \approx (4\pi|x - y|)^{-1}$, i.e. the oscillations of f are locally damped in f_c. Therefore tensor interpolation can be applied to approximate f_c instead of f on suitable axis-parallel boxes t and s and we get

$$f_c(x,y) \approx \sum_{\nu \in M} \sum_{\mu \in M} f_c(\xi_{t,\nu}, \xi_{s,\mu}) L_{t,\nu}^{(m)}(x) L_{s,\mu}^{(m)}(y), \tag{4}$$

where ν and μ are multi-indices in the set $M = \{1, \ldots, m+1\}^3$, $\xi_{t,\nu}$ are tensor products of 1D Chebyshev nodes of order $m+1$ transformed to the box $t = (a_1, b_1] \times \ldots \times (a_3, b_3]$, i.e. $\xi_{t,\nu} = (\xi_{[a_1,b_1],\nu_1}, \xi_{[a_2,b_2],\nu_2}, \xi_{[a_3,b_3],\nu_3})$ with

$$\xi_{[a_j,b_j],\nu_j} = \frac{a_j + b_j}{2} + \frac{b_j - a_j}{2} \cos\left(\frac{2\nu_j - 1}{2\pi(m+1)}\right), \quad \nu_j \in \{1, \ldots, m+1\},$$

and $L_{t,\nu}^{(m)}$ are the corresponding Lagrange polynomials, which are tensor products of the 1D Lagrange polynomials corresponding to the interpolation nodes $\{\xi_{[a_j,b_j],\nu_j}\}_{\nu_j=1}^{m+1}$.

Inserting approximation (4) into (2) and grouping the terms depending on x and y, respectively, yields the desired separable approximation

$$f(x,y) \approx \sum_{\nu \in M} \sum_{\mu \in M} f_c(\xi_{t,\nu}, \xi_{s,\mu}) L_{t,c,\nu}^{(m)}(x) \overline{L_{s,c,\mu}^{(m)}(y)}, \tag{5}$$

$$L_{t,c,\nu}^{(m)}(x) := L_{t,\nu}^{(m)}(x) \exp(i\kappa\langle x, c\rangle). \tag{6}$$

The directional approximation (5) of f can be used to approximate the submatrix $A|_{\hat{t} \times \hat{s}}$ of the matrix A in (1) restricted to the entries of the index sets $\hat{t}$ and $\hat{s}$ for two suitable axis-parallel boxes t and s, i.e.,

$$A|_{\hat{t} \times \hat{s}}[j,k] = f(x_j, y_k) \approx \sum_{\nu \in M} \sum_{\mu \in M} f_c(\xi_{t,\nu}, \xi_{s,\mu}) L_{t,c,\nu}^{(m)}(x_j) \overline{L_{s,c,\mu}^{(m)}(y_k)}. \tag{7}$$

In matrix notation this reads

$$A|_{\hat{t} \times \hat{s}} \approx L_{t,c} A_{c,t \times s} L_{s,c}^*, \tag{8}$$

where we define the *coupling matrix* $A_{c,t \times s} \in \mathbb{C}^{(m+1)^3 \times (m+1)^3}$ by

$$A_{c,t \times s}[j,k] := f_c(\xi_{t,\alpha_j}, \xi_{s,\beta_k}), \quad j,k \in \{1, \ldots, (m+1)^3\}, \tag{9}$$

for suitably ordered multi-indices $\alpha_j, \beta_k \in M = \{1, \ldots, m+1\}^3$, the *directional interpolation matrix* $L_{t,c} \in \mathbb{C}^{\hat{t} \times (m+1)^3}$ by

$$L_{t,c}[j,k] := L_{t,c,\alpha_k}^{(m)}(x_j), \quad j \in \hat{t}, k \in \{1, \ldots, (m+1)^3\}, \tag{10}$$

and $L_{s,c}$ analogously. In particular, instead of the original $\#\hat{t} \cdot \#\hat{s}$ matrix entries only $(m+1)^3(\#\hat{t} + \#\hat{s} + (m+1)^3)$ entries have to be computed for the approximation in (8), which is significantly less if $(m+1)^3 \ll \#\hat{t}, \#\hat{s}$.

In the following admissibility conditions we will specify for which boxes t and s and which direction c the approximation in (5) is applicable. Similar criteria have been considered in [1,5,13]. In particular, the criteria lead to exponential convergence of the approximation with respect to the interpolation degree [5,17].

Definition 1 Directional admissibility [5, cf. Sect. 3.3]). *Let $t, s \subset \mathbb{R}^3$ be two axis-parallel boxes and let $c \in \mathbb{R}^3$ be a direction with $|c| = 1$ or $c = 0$. Denote the midpoints of t and s by m_t and m_s, respectively. Let two constants $\eta_1 > 0$ and $\eta_2 > 0$ be chosen suitably. Define the diameter $\mathrm{diam}\,(t)$ and the distance $\mathrm{dist}\,(t, s)$ by*

$$\mathrm{diam}\,(t) := \sup_{x_1, x_2 \in t} |x_1 - x_2|, \quad \mathrm{dist}\,(t, s) := \inf_{x \in t, y \in s} |x - y|.$$

We say that t and s are directionally admissible with respect to c if the separation criterion

$$\max\{\mathrm{diam}\,(t), \mathrm{diam}\,(s)\} \leq \eta_2 \, \mathrm{dist}\,(t, s), \tag{A1}$$

and the two cone admissibility criteria

$$\kappa \left| \frac{m_t - m_s}{|m_t - m_s|} - c \right| \leq \frac{\eta_1}{\max\{\mathrm{diam}\,(t), \mathrm{diam}\,(s)\}}, \tag{A2}$$

$$\kappa \max\{\mathrm{diam}\,(t), \mathrm{diam}\,(s)\}^2 \leq \eta_2 \, \mathrm{dist}\,(t, s) \tag{A3}$$

are satisfied.

Criterion (A1) is a standard separation criterion, see, e.g., [10] and [11, Sect. 4.2.3]. It ensures that the boxes t and s are well-separated allowing for an approximation of general non-oscillating kernels.

Criterion (A3) is similar to (A1), since it also controls the distance of two boxes t and s. Note that (A1) follows immediately from (A3) in case that $\kappa \max\{\mathrm{diam}\,(t), \mathrm{diam}\,(s)\} > 1$ and vice versa in the opposite case. As stated in [3, Sect. 3], (A3) can also be understood as a bound on the angle between all vectors $x - y$ for $x \in t$ and $y \in s$ that shrinks if κ or $\max\{\mathrm{diam}\,(t), \mathrm{diam}\,(s)\}$ increases. Hence, (A3) guarantees that the angle between $x - y$ and a direction c is small if the angle between the difference of the midpoints $m_t - m_s$ and c is already small, which is enforced by (A2).

Indeed, criterion (A2) is used to assign a suitable direction c to two non-overlapping boxes t and s. While the choice $c = (m_t - m_s)/|m_t - m_s|$ would always guarantee (A2), we want to choose c from a small, finite set of directions. This allows to use the same direction c for a fixed box t and several boxes s_j and, therefore, to use the same interpolation matrix $L_{t,c}$ for the approximation of various blocks $A|_{\hat{t} \times \hat{s}_j}$ as in (8). A possible way to construct suitable sets of directions and further details on criterion (A2) are discussed in Sect. 2.4. First, we want to discuss how to use criteria (A1) and (A3) to construct a suitable partition of the matrix A in (1) based on the clustering described in Sect. 2.1.

2.3   Partitioning of the Matrix

In general, the sets of evaluation points P_T and P_S for the matrix A in (1) are contained in overlapping boxes T and S. Therefore, the full matrix A cannot be

---

Algorithm 2. Construction of a block tree $\mathcal{T}_{T\times S}$

1: **input:** Box cluster trees $\mathcal{T}_T$ and $\mathcal{T}_S$, parameter η_2 for the criteria (A1) and (A3).
2: Set $b = (t_1^0, s_1^0)$, i.e. the pair of roots of $\mathcal{T}_T$ and $\mathcal{T}_S$.
3: Construct an empty tree $\mathcal{T}_{T\times S}$ and add b as its root.
4: Call REFINEBLOCK(b, $\mathcal{T}_{T\times S}$).

5: **function** REFINEBLOCK($b = (t, s)$, $\mathcal{T}_{T\times S}$)
6:      **if** $t \in \mathcal{L}_T$ or $s \in \mathcal{L}_S$ **then**
7:          **return**
8:      **if** t and s violate (A1) or (A3) **then**
9:          **for** $t' \in$ child(t) **do**
10:            **for** $s' \in$ child(s) **do**
11:                Add $b' = (t', s')$ to $\mathcal{T}_{T\times S}$ as child of b.
12:                Call REFINEBLOCK(b', $\mathcal{T}_{T\times S}$).

---

approximated directly. For this reason, we recursively construct a partition of A by Algorithm 2, which we organize in a block tree $\mathcal{T}_{T\times S}$ ([11, Sect. 5.5]).

Definition 2. *Let $\mathcal{T}_T$ and $\mathcal{T}_S$ be two uniform box cluster trees and let $\eta_2 > 0$. A block tree $\mathcal{T}_{T\times S}$ is constructed by Algorithm 2. The set of all leaves of $\mathcal{T}_{T\times S}$ is denoted by $\mathcal{L}_{T\times S}$ and split into the set of admissible (i.e. approximable) leaves and the set of inadmissible leaves*

$$\mathcal{L}_{T\times S}^+ := \{b = (t, s) \in \mathcal{L}_{T\times S} : t \text{ and } s \text{ satisfy (A1) and (A3)}\},$$
$$\mathcal{L}_{T\times S}^- := \mathcal{L}_{T\times S} \setminus \mathcal{L}_{T\times S}^+.$$

For a given block tree $\mathcal{T}_{T\times S}$ the pairs of indices $\hat{t} \times \hat{s}$ of all leaves $(t, s) \in \mathcal{L}_{T\times S}$ form a partition of the full index set $\{1, \ldots, N_T\} \times \{1, \ldots, N_S\}$, i.e. of the matrix A. The matrix blocks corresponding to admissible blocks $b \in \mathcal{L}_{T\times S}^+$ can be approximated by the directional interpolation (8). Inadmissible blocks related to $b \in \mathcal{L}_{T\times S}^-$ are computed directly.

2.4 Choice of Directions

As we would like to use relatively small numbers of directions c in the directional approximations (8), we consider a fixed set of directions $D^{(\ell)}$ for all blocks (t, s) at a given level ℓ of the block tree. These sets $D^{(\ell)}$ should be constructed in such a way that for all blocks (t, s) at level ℓ in $\mathcal{L}_{T\times S}^+$ there exists a direction $c \in D^{(\ell)}$ such that criterion (A2) holds for some fixed η_1.

Since the bound on the right-hand side of (A2) increases for decreasing diameters of t and s and these diameters are halved for each new level of the uniform box cluster trees, the number of directions in $D^{(\ell)}$ can be reduced with increasing level ℓ. If the maximum of the diameters of two boxes t and s at level $\tilde{\ell}$ is so small that the bound on the right-hand side of (A2) is greater than κ, then (A2) holds for $c = 0$ for all following levels. In this case, a plane wave term is

---

Algorithm 3. Construction of directions $D^{(\ell)}$

1: **input**: Largest high frequency level $\ell_{\text{hf}} \geq -1$.
2: **for** $\ell = \ell_{\text{hf}} + 1, \ell_{\text{hf}} + 2, \ldots, \min\{p(\mathcal{T}_T), p(\mathcal{T}_S)\}$ **do**
3:     Set $D^{(\ell)} = \{0\}$.
4: Construct the six faces $\{E_j^{(\ell_{\text{hf}})}\}_{j=1}^6$ of the cube $[-1, 1]^3$, i.e.
  $E_1^{(\ell_{\text{hf}})} = \{-1\} \times [-1, 1]^2, \ E_2^{(\ell_{\text{hf}})} = \{1\} \times [-1, 1]^2, \ \ldots, \ E_6^{(\ell_{\text{hf}})} = [-1, 1]^2 \times \{1\}.$
5: Set $D^{(\ell_{\text{hf}})} = \{c_j^{(\ell_{\text{hf}})}\}_{j=1}^6$ where $c_j^{(\ell_{\text{hf}})}$ is the midpoint of $E_j^{(\ell_{\text{hf}})}$, i.e.
  $c_1^{(\ell_{\text{hf}})} = (-1, 0, 0), \ c_2^{(\ell_{\text{hf}})} = (1, 0, 0), \ \ldots, \ c_6^{(\ell_{\text{hf}})} = (0, 0, 1).$
6: **for** $\ell = \ell_{\text{hf}} - 1, \ldots, 0$ **do**
7:     Set $D^{(\ell)} = \emptyset$.
8:     **for all faces** $E_j^{(\ell+1)}, \ j = 1, \ldots, 6 \cdot 4^{\ell_{\text{hf}}-\ell-1}$ **do**
9:         Uniformly subdivide $E_j^{(\ell+1)}$ into 4 faces $E_{4(j-1)+1}^{(\ell)}, \ \ldots, \ E_{4j}^{(\ell)}$.
10:        Construct normalized midpoints $c_{4(j-1)+1}^{(\ell)}, \ \ldots, \ c_{4j}^{(\ell)}$ of $E_{4(j-1)+1}^{(\ell)}, \ \ldots, \ E_{4j}^{(\ell)}$.
11:        Add directions $c_{4(j-1)+1}^{(\ell)}, \ \ldots, \ c_{4j}^{(\ell)}$ to $D^{(\ell)}$.

---

not needed for the approximation of the Helmholtz kernel f, and the approximation (8) coincides with a standard tensor interpolation. We call the other levels satisfying

$$\frac{\eta_1}{\kappa \max\{\text{diam}(t), \text{diam}(s)\}} \leq 1, \quad \text{for all } t \in \mathcal{T}_T^\ell, s \in \mathcal{T}_S^\ell, \tag{11}$$

high frequency levels and denote the *largest high frequency level* as ℓ_{hf}, or set $\ell_{\text{hf}} = -1$ in case that all levels $\ell \geq 0$ are low frequency levels, i.e. do not satisfy (11). The value of ℓ_{hf} depends on η_1 and the uniform box cluster trees $\mathcal{T}_T$ and $\mathcal{T}_S$. In practice, we choose a suitable level ℓ_{hf} instead of η_1 and construct the *sets of directions* $D^{(\ell)}$, using more and more directions for levels $\ell < \ell_{\text{hf}}$. Our construction by Algorithm 3 combines ideas from [9, Sect. 4.1] and [3, Sect. 3].

Finally, we assign a direction $c \in D^{(\ell)}$ to a pair of boxes t and s which is close to the normalized difference $(m_t - m_s)/|m_t - m_s|$ of the midpoints of t and s and, hence, can be used for the directional approximation (8). For this purpose, we define a mapping $\text{dir}_{(\ell)}$ for each level $\ell \in \mathbb{N}_0$, which maps a vector v in $\mathbb{R}^3 \backslash \{0\}$ to a direction $c_j^{(\ell)}$ such that the intersection point of the ray $\{\lambda v : \lambda > 0\}$ and the surface of the cube $[-1, 1]^3$ lies in the face $E_j^{(\ell)}$ (cf. Algorithm 3).

Definition 3. *Let $\ell_{\text{hf}} \geq -1$ and let the directions $D^{(\ell)}$ and the faces $\{E_j^{(\ell)}\}$ be constructed by Algorithm 3. We define the mapping $\text{dir}_{(\ell)} : \mathbb{R}^3 \to D^{(\ell)} \cup \{0\}$ for each $\ell \in \mathbb{N}_0$ as follows:*

- *If $\ell > \ell_{\text{hf}}$ we set $\text{dir}_{(\ell)}(v) = 0$ for all $v \in \mathbb{R}^3$.*
- *If $\ell \leq \ell_{\text{hf}}$ we set $\text{dir}_{(\ell)}(0) = 0$. For all $v \in \mathbb{R}^3 \backslash \{0\}$ we set $\text{dir}_{(\ell)}(v) = c_{j(v)}^{(\ell)}$ where*

$$j(v) := \min\{j : \psi_Q(v) \in E_j^{(\ell)}\}, \quad \psi_Q(v) := \frac{1}{\max_{j \in \{1, \ldots, 3\}} |v_j|} v$$

to avoid ambiguity.

For two boxes $t, s \subset \mathbb{R}^3$ and a level $\ell \in \mathbb{N}_0$ we define the direction $c_{(\ell)}(t, s)$ by

$$c_{(\ell)}(t, s) := \text{dir}_{(\ell)}\left(\frac{m_t - m_s}{|m_t - m_s|}\right).$$

In this way (A2) is satisfied for two boxes t, s, and the direction $c_{(\ell)}(t, s)$ for a constant η_1 which depends linearly on the product $\kappa q_{\ell_{\text{hf}}}$ [17, Theorem. 2.19]. Here $q_{\ell_{\text{hf}}}$ denotes the maximal diameter of all boxes at level ℓ_{hf} in the trees $\mathcal{T}_T$ and $\mathcal{T}_S$.

2.5   Transfer Operations

The approximation of an admissible subblock $A|_{\hat{t} \times \hat{s}}$ of A in (7) can be further enhanced. If t is a non-leaf box at level ℓ in a box cluster tree $\mathcal{T}_T$ with children $t_1, \ldots, t_k$, the directional interpolation matrix $L_{t,c}$ can be approximated using the matrices $L_{t_j, c_{\ell+1}}$ for a suitable direction $c_{\ell+1}$. We describe this approach following [5, Sect. 2.2.2].

Let us rewrite the generating functions $L_{t,c,\nu}^{(m)}(x)$ of $L_{t,c}$ in (6) by

$$L_{t,c,\nu}^{(m)}(x) = \exp(i\kappa\langle x, c_{\ell+1}\rangle)\left[\exp(i\kappa\langle x, c - c_{\ell+1}\rangle)L_{t,\nu}^{(m)}(x)\right].$$

If $c_{\ell+1}$ is sufficiently close to c, the term in square brackets is smooth and can be interpolated for points x in a child box t_j yielding

$$\exp(i\kappa\langle x, c - c_{\ell+1}\rangle)L_{t,\nu}^{(m)}(x) \approx \sum_{\tilde{\nu} \in M} \exp(i\kappa\langle \xi_{t_j, \tilde{\nu}}, c - c_{\ell+1}\rangle)L_{t,\nu}^{(m)}(\xi_{t_j, \tilde{\nu}})L_{t_j, \tilde{\nu}}^{(m)}(x).$$

This provides an approximation of the restriction of $L_{t,c,\nu}^{(m)}$ to the child t_j

$$L_{t,c,\nu}^{(m)}\big|_{t_j}(x) \approx \sum_{\tilde{\nu} \in M} \left(\left[\exp(i\kappa\langle \xi_{t_j, \tilde{\nu}}, c - c_{\ell+1}\rangle)L_{t,\nu}^{(m)}(\xi_{t_j, \tilde{\nu}})\right] L_{t_j, c_{\ell+1}, \tilde{\nu}}^{(m)}(x)\right).$$

In matrix notation the related restriction to the index set $\hat{t}_j$ reads as

$$L_{t,c}|_{\hat{t}_j \times (m+1)^3} \approx L_{t_j, c_{\ell+1}} E_{t_j, c}, \tag{12}$$

where the entries of the transfer matrix $E_{t_j, c} \in \mathbb{C}^{(m+1)^3 \times (m+1)^3}$ are defined by

$$E_{t_j, c}[k, \ell] := \exp(i\kappa\langle \xi_{t_j, \nu_k}, c - c_{\ell+1}\rangle)L_{t, \nu_\ell}^{(m)}(\xi_{t_j, \nu_k}), \tag{13}$$

for all $k, \ell \in \{1, \ldots, (m+1)^3\}$.

A suitable choice [17, Theorem. 2.19] for the direction $c_{\ell+1}$ is given by $\text{dir}_{(\ell+1)}(c)$, with $\text{dir}_{(\ell+1)}$ given in Definition 3. Since this direction depends only on c and the level ℓ of the box t, it is reasonable to omit the dependence of the transfer matrix $E_{t_j, c}$ on $c_{\ell+1}$ in the notation.

2.6 Main Algorithm

In the previous sections, we have described how to partition the matrix (1) and how to approximate suitable subblocks. Here we explain the complete algorithm for a a matrix-vector multiplication $g = Av$.

The idea is to execute the multiplication blockwise according to the partition induced by the leaves $\mathcal{L}_{T \times S}$ of the block tree $\mathcal{T}_{T \times S}$. Inadmissible blocks from $\mathcal{L}_{T \times S}^-$ are multiplied directly with the target vector v. For admissible blocks from $\mathcal{L}_{T \times S}^+$ we use the decomposition (8) and split the multiplication into three phases. This is similar to the usual three-phase algorithm for $\mathcal{H}^2$-matrices [11, Sect. 8.7] and the FMM [10] with adaptations due to the directional approximation. We describe the scheme first for one block corresponding to an admissible pair of boxes t and s at level ℓ and then give a description of the complete algorithm.

In the first phase, the forward transformation, the product $\tilde{v}_{s,c} := L_{s,c}^* v|_{\hat{s}}$ is computed. If s is a leaf in the cluster tree, this is done directly by (10). This is also known as S2M (source to moment) step in fast multipole methods. If s is not a leaf, approximation (12) with $c_{\ell+1} = \mathrm{dir}_{(\ell+1)}(c)$ is used iteratively to get

$$\tilde{v}_{s,c} := \sum_{s_j \in \mathrm{child}(s)} E_{s_j,c}^* \tilde{v}_{s_j,c_{\ell+1}} \approx \sum_{s_j \in \mathrm{child}(s)} E_{s_j,c}^* \left[L_{s_j,c_{\ell+1}}^* v|_{\hat{s}_j} \right] \approx L_{s,c}^* v|_{\hat{s}},$$

by using the products of the children, which is also known as M2M operation (moment to moment). In the second phase, which is called multiplication phase or M2L (moment to local) step, the product

$$\tilde{g}_{t,c} := A_{c,t \times s} \tilde{v}_{s,c}$$

is computed by (9). In the complete algorithm all contributions from various boxes s are added up, i.e.

$$\tilde{g}_{t,c} := \sum_{s:(t,s) \in \mathcal{L}_{T \times S}^+} A_{c,t \times s} \tilde{v}_{s,c}.$$

In the third phase, the so-called backward transformation, the product

$$\tilde{g}|_{\hat{t}} := L_{t,c} \tilde{g}_{t,c} \tag{14}$$

is computed. If t is a leaf, this is done directly. This step is known as L2T (local to target) in fast multipole methods. If t is not a leaf, the approximation (12) is used to compute

$$\tilde{g}_{t_j,c_{\ell+1}} = E_{t_j,c} \, \tilde{g}_{t,c}, \tag{15}$$

for all children t_j of t, which is also known as L2L operation (local to local), and the evaluation (14) takes place for descendants which are leaves. In the complete algorithm the local contribution in (15) is added to the existing contribution $\tilde{g}_{t_j,c_{\ell+1}}$ originating from the multiplication phase.

Before we present the complete Algorithm 4, we define the sets of active and inherited directions for each box in the cluster trees $\mathcal{T}_T$ and $\mathcal{T}_S$. These are used to keep track of all required directions for boxes t and s in the trees $\mathcal{T}_T$ and $\mathcal{T}_S$. They can be generated during the construction of the block tree $\mathcal{T}_{T \times S}$.

Definition 4. *Let $\mathcal{T}_T$ and $\mathcal{T}_S$ be two uniform box cluster trees, $\mathcal{T}_{T \times S}$ the corresponding block tree and $\ell_{\mathrm{hf}} \geq -1$. Recalling Algorithm 3 and Definition 3 we define for all $\ell \geq 0$ and all $t \in \mathcal{T}_T^{(\ell)}$ the set of active directions by*

$$D(t) := \{c \in D^{(\ell)} : \exists\, s \in \mathcal{T}_S^{(\ell)} \text{ such that } (t,s) \in \mathcal{L}_{T \times S}^{+} \text{ and } c = c_{(\ell)}(t,s)\}.$$

The set of inherited directions $\hat{D}(t)$ is defined recursively by setting $\hat{D}(t_1^0) = \emptyset$ for the root t_1^0 of $\mathcal{T}_T$, and for all $\ell > 0$ and all $t \in \mathcal{T}_T^{(\ell)}$ by setting

$$\hat{D}(t) := \{\hat{c} \in D^{(\ell)} : \exists\, c \in D(t') \cup \hat{D}(t') \text{ such that } \hat{c} = \mathrm{dir}_{(\ell)}(c),\ t' = \mathrm{parent}(t)\}.$$

Analogously, the sets of active directions $D(s)$ and inherited directions $\hat{D}(s)$ are defined for clusters $s \in \mathcal{T}_S$.

Algorithm 4. Fast directional matrix vector multiplication $g \approx Av$

1: **input**: Box cluster trees $\mathcal{T}_T$ and $\mathcal{T}_S$, block tree $\mathcal{T}_{T \times S}$, interpolation degree m,
    sets of directions $D(t)$, $\hat{D}(t)$, $D(s)$, $\hat{D}(s)$ for all boxes t, s in $\mathcal{T}_T$, $\mathcal{T}_S$.
2: Initialize $g = 0$.
3: ▷ *Forward transformation*
4: **for** all leaves $s \in \mathcal{L}_S$ **do**
5:     **for** all directions $c \in D(s) \cup \hat{D}(s)$ **do**
6:         Compute $\tilde{v}_{s,c} = L_{s,c}^{*} v|_{\hat{s}}$.
7: **for** all levels $\ell = p(\mathcal{T}_S) - 1, \ldots, 0$ **do**
8:     **for** all non-leaf boxes $s \in \mathcal{T}_S^{(\ell)} \backslash \mathcal{L}_S$ **do**
9:         **for** all directions $c \in D(s) \cup \hat{D}(s)$ **do**
10:             Set $\tilde{v}_{s,c} = 0$.
11:             **for** all $s' \in \mathrm{child}(s)$ **do**
12:                 Update $\tilde{v}_{s,c} \mathrel{+}= E_{s',c}^{*} \tilde{v}_{s',c'}$, where $c' = \mathrm{dir}_{(\ell+1)}(c)$.
13: ▷ *Multiplication phase*
14: **for** all boxes $t \in \mathcal{T}_T$ **do**
15:     **for** all directions $c \in D(t) \cup \hat{D}(t)$ **do**
16:         Initialize $\tilde{g}_{t,c} = 0$.
17:     **for** all boxes $s \in \mathcal{T}_S$ such that $(t,s) \in \mathcal{L}_{T \times S}^{+}$ **do**
18:         Update $\tilde{g}_{t,c} \mathrel{+}= A_{c,t \times s} \tilde{v}_{s,c}$, where $c = \mathrm{dir}_{(\ell)}(t,s)$ and $\ell = \mathrm{level}(t)$.
19: ▷ *Backward transformation*
20: **for** all levels $\ell = 0, \ldots, p(\mathcal{T}_T) - 1$ **do**
21:     **for** all non-leaf boxes $t \in \mathcal{T}_T^{(\ell)} \backslash \mathcal{L}_T$ **do**
22:         **for** all directions $c \in D(t) \cup \hat{D}(t)$ **do**
23:             **for** all $t' \in \mathrm{child}(t)$ **do**
24:                 Update $\tilde{g}_{t',c'} \mathrel{+}= E_{t',c} \tilde{g}_{t,c}$, where $c' = \mathrm{dir}_{(\ell+1)}(c)$.
25: **for** all leaves $t \in \mathcal{L}_T$ **do**
26:     **for** all directions $c \in D(t) \cup \hat{D}(t)$ **do**
27:         Update $g|_{\hat{t}} \mathrel{+}= L_{t,c} \tilde{g}_{t,c}$.
28: ▷ *Nearfield evaluation*
29: **for** all blocks $b = (t,s) \in \mathcal{L}_{T \times S}^{-}$ **do**
30:     Update $g|_{\hat{t}} \mathrel{+}= A|_{\hat{t} \times \hat{s}} v|_{\hat{s}}$.

2.7   Implementation Details

In this section, we describe how to exploit the uniformity of the box cluster trees to reduce the storage required by the transfer matrices $E_{t',c}$ defined in (13) and the coupling matrices $A_{c,t\times s}$ defined in (9). This is crucial as there is a large number of such matrices involved in the computations in Algorithm 4.

For a level $\ell \geq 0$, a box $t \in \mathcal{T}_T^\ell$, a child t' and directions c and $c' = \mathrm{dir}_{(\ell+1)}(c)$ we consider the transfer matrix $E_{t',c}$ which has the entries

$$E_{t',c}[j,k] = \exp(i\kappa\langle \xi_{t',\nu_j}, c - c'\rangle)L_{t,\nu_k}^{(m)}(\xi_{t',\nu_j}), \quad j,k \in \{1,\ldots,(m+1)^3\}.$$

This matrix can be split into a directional and a non-directional part by

$$E_{t',c} = E_{t',c}^{\mathrm{d}}E_{t'},$$

where we define the directional part $E_{t',c}^{\mathrm{d}}$ and the non-directional part $E_{t'}$ by

$$E_{t',c}^{\mathrm{d}} := \mathrm{diag}\left(\{\exp(i\kappa\langle \xi_{t',\tilde{\nu}}, c - c'\rangle)\}_{\tilde{\nu}\in M}\right),$$
$$E_{t'}[j,k] := L_{t,\nu_k}^{(m)}(\xi_{t',\nu_j}), \quad j,k \in \{1,\ldots,(m+1)^3\}.$$

Let us consider the non-directional part $E_{t'}$ first. The value of the Lagrange polynomial $L_{t,\nu}^{(m)}$ depends only on the position of the evaluation point $\xi_{t',\mu}$ relative to the box t. Together with the uniformity of the box cluster tree $\mathcal{T}_T$, this implies that each $E_{t'}$ is identical to one of 8 non-directional transfer matrices in a reference configuration. Only these reference matrices of size $(m+1)^3 \times (m+1)^3$ have to be computed and stored. The directional part $E_{t',c}^{\mathrm{d}}$ changes for varying boxes t, child boxes t' or directions c. Since it is diagonal, however, only $(m+1)^3$ entries instead of $(m+1)^6$ entries need to be computed. Furthermore, for low frequency levels $\ell > \ell_{\mathrm{hf}}$ the directional part $E_{t',c}^{\mathrm{d}}$ becomes the identity and no additional computations are required.

Next we consider the coupling matrices $A_{c,t\times s}$ defined in (9) for admissible blocks (t,s) in a block tree $\mathcal{T}_{T\times S}$. $A_{c,t\times s}$ depends on the difference of the cluster centers only, see (3). Due to the uniformity of the box cluster trees, many of the coupling matrices coincide. In particular, it suffices to compute and store all required coupling matrices for all levels ℓ only once for a reference configuration and assign them to the appropriate blocks $(t,s) \in \mathcal{L}_{T\times S}^+$.

The dimension of the coupling matrices (9) increases cubically in the interpolation degree m. A compression of these matrices by a low rank approximation

$$A_{c,t\times s} \approx U_{c,t\times s}V_{c,t\times s}^*,$$

with $U_{c,t\times s}, V_{c,t\times s} \in \mathbb{C}^{(m+1)^3 \times k}$ for some low rank k, increases the performance of the algorithm (cf. [13]). Such approximations exist because the coupling matrices are generated by smooth functions. For their construction, we apply a partially pivoted ACA [2,15] in our implementation and the examples in Sect. 4, but do not analyze its effect on the complexity in the following section. A more involved compression strategy is described in [4].

3   Complexity Analysis

To analyze the complexity of Algorithm 4 for fast directional matrix vector multiplications, we estimate the number of directional interpolation matrices and transfer matrices in Theorem 1, give then an estimate for the number of coupling matrices in Theorem 2 and 3 and finally estimate the number of nearfield matrices in Theorem 4. We start by establishing the general setting.

Throughout this section we fix the wave number $\kappa > 0$ and the sets of points $P_T = \{x_j\}_{j=1}^{N_T}$ and $P_S = \{y_k\}_{k=1}^{N_S}$, which may but do not have to coincide, and set $N = \max\{N_T, N_S\}$. In all considerations $\mathcal{T}_T$ and $\mathcal{T}_S$ denote two uniform box cluster trees as constructed in Algorithm 1 for a fixed parameter $n_{\max}$. We set the maximum and the minimum of the depths of the trees $\mathcal{T}_T$ and $\mathcal{T}_S$

$$p_{\max} := \max\{p(\mathcal{T}_T), p(\mathcal{T}_S)\}, \quad p_{\min} := \min\{p(\mathcal{T}_T), p(\mathcal{T}_S)\}.$$

The diameters of all boxes at a fixed level ℓ of $\mathcal{T}_T$ are identical and denoted as $q_\ell(\mathcal{T}_T)$ just like the diameters $q_\ell(\mathcal{T}_S)$ of boxes at level ℓ in $\mathcal{T}_S$. For all levels $\ell \leq p_{\min}$ we define

$$q_\ell := \max\{q_\ell(\mathcal{T}_T), q_\ell(\mathcal{T}_S)\}.$$

The related block tree $\mathcal{T}_{T \times S}$ is constructed by Algorithm 2 for a fixed parameter η_2. For the directional approximation we use a small, fixed interpolation degree m and the directions $D^{(\ell)}$, constructed by Algorithm 3 for a fixed choice of the largest high frequency level $\ell_{\mathrm{hf}} \geq -1$.

For the complexity analysis we will need a few assumptions which we collect and discuss here. We assume that there exist small constants c_{geo}, $c_{\max}$, c_{ad} and $c_{\mathrm{un}} \in \mathbb{R}_+$ such that the following assumptions hold true:

$$n_{\max} \leq c_{\max}(m+1)^3, \tag{16}$$

$$q_0 \leq c_{\mathrm{un}} \min\{q_0(\mathcal{T}_T), q_0(\mathcal{T}_S)\}, \tag{17}$$

$$p_{\max} \leq \log_8(N) + c_{\mathrm{ad}}, \tag{18}$$

$$\kappa q_0 \leq c_{\mathrm{geo}} \sqrt[3]{N}. \tag{19}$$

In addition, ℓ_{hf} is assumed to be chosen such that

$$\ell_{\mathrm{hf}} + 1 \leq p_{\max} + c_{\mathrm{hf}}, \tag{20}$$

for a small constant $c_{\mathrm{hf}} \in \mathbb{N}_0$. Furthermore, we require that (33) holds, which we introduce and discuss later. Let us shortly discuss above assumptions. By equation (16) we ensure that the maximal number of points in leaf boxes of the cluster trees is reasonably small. Assumption (17) means that the diameters of the root boxes of $\mathcal{T}_T$ and $\mathcal{T}_S$ should be of comparable size. While this is not satisfied in general, one can enforce it by an initial subdivision of the greater box and application of the method to the resulting subboxes. Equation (18) is an indirect assumption on the sets of points P_T and P_S, which holds if the points are distributed more or less uniformly in a 3D domain. Also Eq. (19) is reasonable only if points are distributed rather uniformly in a 3D volume, and guarantees that the wave length $\lambda = 2\pi/\kappa$ is resolved in that case, which is required in typical physical applications. Finally, Eq. (20) is a bound on the largest high frequency level ℓ_{hf} and allows to bound the number of directions constructed in Algorithm 3. With these assumptions we can start with the complexity analysis, which is based on the following obvious, but important observation.

Remark 1. In Algorithm 4 every directional interpolation matrix $L_{t,c}$ and $L_{s,c}$, every transfer matrix $E_{t',c}$ and $E_{s',c}$, every coupling matrix $A_{c,t\times s}$ and every nearfield matrix $A|_{\hat{t}\times\hat{s}}$ is multiplied with a suitable vector exactly once. All entries of these matrices can be computed with $\mathcal{O}(1)$ operations. Since the complexity of the application of a matrix to a vector is proportional to the number of its entries, it suffices to count all these matrices and their respective entries to estimate the storage and runtime complexity of Algorithm 4.

Theorem 1. *Let assumption (20) hold true. Then there exists a constant c_{LE} depending only on c_{hf} such that the number N_{LE} of applied transfer matrices $E_{t',c}$ and $E_{s',c}$ and directional interpolation matrices $L_{t,c}$ and $L_{s,c}$ in Algorithm 4 is bounded by*

$$N_{\mathrm{LE}} \leq c_{\mathrm{LE}}\, 8^{p_{\max}}. \tag{21}$$

If (16) and (18) apply in addition, these matrices can be stored and applied with complexity $\mathcal{O}(N)$.

Proof. We start to estimate the number $N_{\mathrm{LE},T}$ of applied transfer matrices $E_{t',c}$ for L2L operations in lines 20–24 of Algorithm 4 and directional interpolation matrices $L_{t,c}$ for L2T operations in lines 25–27. For this purpose we estimate the number of such matrices for each box t in $\mathcal{T}_T$.

Let us first assume, that $t \in \mathcal{T}_T^\ell$ is a non-leaf box at level ℓ. In this case a transfer matrix is applied for each direction $c \in D(t) \cup \hat{D}(t)$ and each box $t' \in \mathrm{child}(t)$, but no directional interpolation matrix. The number of directions in $D(t) \cup \hat{D}(t)$ is bounded by $\#D^{(\ell)}$, which is $6 \cdot 4^{\ell_{\mathrm{hf}}-\ell}$ if $\ell \leq \ell_{\mathrm{hf}}$ and 1 else, and $\#\mathrm{child}(t) \leq 8$ for all t due to the uniformity of the box cluster tree. Therefore, the total number $N_{\mathrm{LE}}(t)$ of transfer and directional interpolation matrices needed for a non-leaf box $t \in \mathcal{T}_T^\ell$ is bounded by

$$B^{(\ell)} = \begin{cases} 48 \cdot 4^{\ell_{\mathrm{hf}}-\ell}, & \text{if } \ell \leq \ell_{\mathrm{hf}}, \\ 8, & \text{otherwise.} \end{cases}$$

If $t \in \mathcal{T}_T^\ell$ is a leaf box then we only need a directional interpolation matrix for each direction $c \in D(t) \cup \hat{D}(t)$ but no transfer matrix. Therefore, $N_{\mathrm{LE}}(t)$ is bounded by $6 \cdot 4^{\ell_{\mathrm{hf}}-\ell}$ if $\ell \leq \ell_{\mathrm{hf}}$ and by 1 otherwise. Since this bound is less than $B^{(\ell)}$ for all levels ℓ, there holds $N_{\mathrm{LE}}(t) \leq B^{(\ell)}$ for all boxes $t \in \mathcal{T}_T^\ell$.

The number $N_{\mathrm{LE},T}$ of all directional interpolation matrices and transfer matrices for boxes $t \in \mathcal{T}_T$ can hence be estimated by

$$N_{\mathrm{LE},T} = \sum_{\ell=0}^{p(\mathcal{T}_T)} \sum_{t\in\mathcal{T}_T^\ell} N_{\mathrm{LE}}(t) \leq \sum_{\ell=0}^{p(\mathcal{T}_T)} \#\mathcal{T}_T^\ell\, B^{(\ell)}.$$

Due to the uniformity of the box cluster tree there holds $\#\mathcal{T}_T^\ell \leq 8^\ell$. Let us first assume that all levels in $\mathcal{T}_T$ are high frequency levels, i.e. $p(\mathcal{T}_T) \leq \ell_{\mathrm{hf}}$. Then we can further estimate

$$N_{\mathrm{LE},T} \leq \sum_{\ell=0}^{p(\mathcal{T}_T)} 48 \cdot 4^{\ell_{\mathrm{hf}}-\ell} 8^\ell < 48 \cdot 4^{\ell_{\mathrm{hf}}} 2^{p(\mathcal{T}_T)+1} \leq 24 \cdot 4^{c_{\mathrm{hf}}} 8^{p_{\max}}, \tag{22}$$

where we used assumption (20) in the last step. If instead $p(\mathcal{T}_T) > \ell_{\mathrm{hf}}$, we get

$$N_{\mathrm{LE},T} \leq \sum_{\ell=0}^{\ell_{\mathrm{hf}}} 48 \cdot 4^{\ell_{\mathrm{hf}}-\ell} 8^\ell + \sum_{\ell=\ell_{\mathrm{hf}}+1}^{p(\mathcal{T}_T)} 8^{\ell+1} \tag{23}$$
$$\leq 12 \cdot 8^{\ell_{\mathrm{hf}}+1} + 8 \left(8^{p(\mathcal{T}_T)+1} - 8^{\ell_{\mathrm{hf}}+1}\right) \leq 68 \cdot 8^{p_{\max}}.$$

Analogously, we can estimate the number $N_{\mathrm{LE},S}$ of transfer matrices and directional interpolation matrices needed for the S2M and M2M operations in lines 4–12 of Algorithm 4. Therefore, the estimate on the number N_{LE} of all transfer and directional interpolation matrices in (21) holds with $c_{\mathrm{LE}} = 2 \cdot \max(68, 24 \cdot 4^{c_{\mathrm{hf}}})$.

To prove the complexity statement we observe that every transfer matrix (13) has $(m+1)^6$ entries and every directional interpolation matrix (10) has at most $n_{\max}(m+1)^3 \leq c_{\max}(m+1)^6$ entries by assumption (16). Therefore, the linear complexity is a direct consequence of (21), if in addition (18) holds. $\square$

Theorem 2. *Let assumption* (17) *hold true. Then there exists a constant* c_{C} *depending only on* c_{un} *and* η_2*, such that the number* N_{C} *of all coupling matrices* $A_{c,t \times s}$ *in Algorithm 4 is bounded by*

$$N_{\mathrm{C}} \leq c_{\mathrm{C}} \left(p_{\min}(q_0\kappa)^3 + 8^{p_{\min}}\right). \tag{24}$$

If in addition (18) *and* (19) *hold true, these matrices can be stored and applied with complexity* $\mathcal{O}(N \log(N))$*. If* (19) *is replaced by the stronger assumption*

$$\kappa q_0 \leq c \sqrt[3]{N/\log(N)}, \tag{25}$$

then the complexity is reduced to $\mathcal{O}(N)$*.*

Proof. In this proof we pursue similar ideas as in [3, cf. proof of Lem. 8]. We assume that the depth of $\mathcal{T}_{T \times S}$ is not zero, because otherwise $N_{\mathrm{C}} \leq 1$ and the assertion is trivial. Our strategy is to estimate the numbers $N_{\mathrm{C}}^{(\ell)}$ of coupling matrices at all relevant levels $\ell = 1, \ldots, p_{\min}$.

In line 17 of Algorithm 4 we see that the number of coupling matrices needed for a box $t \in \mathcal{T}_T^{(\ell)}$ is given by $\#\{s : (t,s) \in \mathcal{L}_{T \times S}^+\}$. For such blocks $(t,s) \in \mathcal{L}_{T \times S}^+$ the parent(s) is in the nearfield $\mathcal{N}(t_{\mathrm{p}})$ of $t_{\mathrm{p}} := \mathrm{parent}(t)$ by construction of the block tree in Algorithm 2, where

$$\mathcal{N}(t_{\mathrm{p}}) := \{s_{\mathrm{p}} \in \mathcal{T}_S^{(\ell-1)} : s_{\mathrm{p}} \text{ and } t_{\mathrm{p}} \text{ violate (A1) or (A3)}\}.$$

Using this property and the uniformity of the box cluster trees we can estimate

$$N_{\mathrm{C}}^{(\ell)} = \sum_{t \in \mathcal{T}_T^{(\ell)}} \sum_{(t,s) \in \mathcal{L}_{S \times T}^+} 1 \leq \sum_{t \in \mathcal{T}_T^{(\ell)}} 8 \cdot \#\mathcal{N}(\mathrm{parent}(t)) \leq 8^{\ell+1} N_{\mathcal{N},T}^{(\ell-1)}, \qquad (26)$$

where $N_{\mathcal{N},T}^{(\ell-1)}$ is an upper bound for the number of boxes in the nearfield of a box at level $\ell - 1$ in $\mathcal{T}_T$ which we estimate in the following.

We cover the nearfield $\mathcal{N}(t)$ of a fixed box $t \in \mathcal{T}_T^{(\ell)}$ by a ball $B_{r_\ell}(m_t)$ with radius r_ℓ and center m_t and take the ratio of the volume of the ball and the one of a box to estimate $N_{\mathcal{N},T}^{(\ell)}$ for $\ell \geq 1$. We have to distinguish the cases of the two admissibility criteria (A1) and (A3). For this purpose, let $\tilde{\ell}$ be such that $\kappa q_j > 1$, if and only if $j \leq \tilde{\ell}$. Such an $\tilde{\ell}$ exists since q_j decreases monotonically for increasing level j. In particular, we set $\tilde{\ell} = -1$, if $\kappa q_j \leq 1$ for all $j \geq 0$. If $j \leq \tilde{\ell}$ criterion (A3) implies (A1) as mentioned in Sect. 2.2. Vice versa, (A1) implies (A3) if $j > \tilde{\ell}$.

Let us first assume that $\ell \leq \tilde{\ell}$ and consider an arbitrary box $s \in \mathcal{N}(t)$. Then t and s violate (A3), which means that $\eta_2 \, \mathrm{dist}\,(t,s) < \kappa q_\ell^2$, i.e. there exist $x \in \bar{t}$ and $y \in \bar{s}$ such that $|x - y| < \kappa q_\ell^2/\eta_2$. Hence, we can estimate

$$\max_{z \in \bar{s}} |z - m_t| \leq \max_{z \in \bar{s}} \left(|z - y| + |x - y| + |x - m_t| \right)$$

$$\leq q_\ell + \frac{\kappa q_\ell^2}{\eta_2} + \frac{q_\ell}{2} \leq \left(\frac{3}{2} + \frac{1}{\eta_2} \right) \kappa q_\ell^2 =: r_\ell, \qquad (27)$$

where we used $\kappa q_\ell > 1$ for the last estimate. Therefore, every box $s \in \mathcal{N}(t)$ is contained in the ball $B_{r_\ell}(m_t)$ with r_ℓ from (27). If instead $\ell > \tilde{\ell}$ we analogously show

$$\mathcal{N}(t) \subset B_{r_\ell}(m_t), \quad r_\ell = \left(\frac{3}{2} + \frac{1}{\eta_2} \right) q_\ell. \qquad (28)$$

With the ball $B_{r_\ell}(m_t)$ covering $\mathcal{N}(t)$ we can estimate

$$\#\mathcal{N}(t) \leq \frac{|B_{r_\ell}(m_t)|}{v_\ell(\mathcal{T}_S)} = \frac{(4\pi/3)r_\ell^3}{3^{-3/2} q_\ell(\mathcal{T}_S)^3} = 4\pi\sqrt{3} \left(\frac{r_\ell}{q_\ell(\mathcal{T}_S)} \right)^3, \qquad (29)$$

where $v_\ell(\mathcal{T}_S) = 3^{-3/2} q_\ell(\mathcal{T}_S)^3$ denotes the volume of boxes $s \in \mathcal{T}_S^{(\ell)}$. Since $t \in \mathcal{T}_T^{(\ell)}$ was arbitrary, the bound in (29) holds also for $N_{\mathcal{N},T}^{(\ell)}$ instead of $\mathcal{N}(t)$.

Summarizing (26) and above findings, we get for the number N_{C} of all coupling matrices the estimate

$$N_{\mathrm{C}} = \sum_{\ell=1}^{p_{\min}} N_{\mathrm{C}}^{(\ell)} \leq \sum_{\ell=1}^{p_{\min}} 8^{\ell+1} N_{\mathcal{N},T}^{(\ell-1)} \leq \sum_{\ell=1}^{p_{\min}} 8^{\ell+1} 4\pi\sqrt{3} \left(\frac{r_{\ell-1}}{q_{\ell-1}(\mathcal{T}_S)} \right)^3$$

$$\leq 4\pi\sqrt{3} \left(\frac{3}{2} + \frac{1}{\eta_2} \right)^3 \left(\sum_{\ell=1}^{\tilde{\ell}+1} 8^{\ell+1} (\kappa c_{\mathrm{un}} q_\ell)^3 + \sum_{\ell=\tilde{\ell}+2}^{p_{\min}} 8^{\ell+1} c_{\mathrm{un}}^3 \right)$$

$$\leq c_{\mathrm{C}} \left(\sum_{\ell=1}^{\tilde{\ell}+1} (\kappa q_0)^3 + 8^{p_{\min}} \right) \leq c_{\mathrm{C}} (p_{\min} (\kappa q_0)^3 + 8^{p_{\min}}), \qquad (30)$$

where we assumed that $1 \leq \tilde{\ell} + 1 < p_{\min}$ and used assumption (17) and the relation $q_0 = 2^{\ell} q_{\ell}$. If either $\tilde{\ell} + 1 \geq p_{\min}$ or $\tilde{\ell} = -1$, one can repeat the estimates in (30) and ends up with a similar result where one can cancel $8^{p_{\min}}$ in the first case and $p_{\min}(\kappa q_0)^3$ in the second case. The assertions about the complexity follow directly from (30) with assumptions (18) and (19) or (25), respectively, since every coupling matrix (9) has $(m+1)^6 = \mathcal{O}(1)$ entries. □

In Theorem 2 we have estimated the number N_C of all coupling matrices, which corresponds to the number of admissible blocks $\mathcal{L}^+_{T \times S}$. As explained in Sect. 2.7, we store reoccuring matrices only once to reduce the related storage costs drastically as we will see in the next theorem and in Sect. 4. Since one needs to know all blocks in $\mathcal{L}^+_{T \times S}$ in Algorithm 4 and storing them has complexity $\mathcal{O}(N_C)$, storing each matrix only once does not reduce the overall storage complexity of the method asymptotically.

Theorem 3. *Let the root boxes T and S of $\mathcal{T}_T$ and $\mathcal{T}_S$ be identical up to translation. Then the number N_{SC} of coupling matrices $A_{c,t \times s}$ which have to be stored can be estimated by*

$$N_{SC} \leq p_{\min} \max\{c_C, c_C^{2/3}(\kappa q_0)^2\}. \tag{31}$$

If (18) and (19) hold, the corresponding storage complexity is $\mathcal{O}(N^{2/3} \log(N))$.

Proof. From the proof of Theorem 2, in particular (27), (28) and (29), it follows that the number of admissible blocks $(t, s) \in \mathcal{L}^+_{T \times S}$ for a fixed box $t \in \mathcal{T}_T^{\ell}$ can be estimated by

$$8 \cdot \#\mathcal{N}(\text{parent}(t)) \leq 32\pi\sqrt{3}c_{\text{un}}^3 \left(\frac{3}{2} + \frac{1}{\eta_2}\right)^3 \max\{1, (\kappa q_0)^3 8^{1-\ell}\} \tag{32}$$
$$\leq c_C \max\{1, (\kappa q_0)^3 8^{-\ell}\},$$

where we used $q_\ell = 2^{-\ell} q_0$, and c_C is the same constant as in (24). For a different box $t' \in \mathcal{T}_T^{\ell}$ the boxes s' such that $(t', s') \in \mathcal{L}^+_{T \times S}$ are identical to blocks (t, s) up to translation, which follows from the assumption on the root boxes T and S and the uniformity of the trees $\mathcal{T}_T$ and $\mathcal{T}_S$. Hence, the coupling matrices coincide and (32) is a bound for the number $N_{SC}^{(\ell)}$ of stored coupling matrices at level ℓ. On the other hand, there are at most $8^{2\ell}$ blocks at level ℓ of $\mathcal{T}^+_{T \times S}$, which gives

$$N_{SC}^{(\ell)} \leq \min\{8^{2\ell}, c_C \max\{1, (\kappa q_0)^3 8^{-\ell}\}\}.$$

The maximum over all ℓ of the expression on the right-hand side is bounded by $8^{2\ell^*}$, where ℓ^* is the intersection point of $8^{2\ell}$ and $c_C \max\{1, (\kappa q_0)^3 8^{-\ell}\}$. By computing this maximum we end up with the general bound

$$N_{SC}^{(\ell)} \leq \max\left\{c_C, c_C^{2/3}(\kappa q_0)^2\right\} \quad \text{for all } \ell \geq 0.$$

Summation over all levels $\ell = 1, \dots, p_{\min}$ yields (31). Since every coupling matrix has $(m+1)^6$ entries, it follows that all distinct coupling matrices can be stored with $\mathcal{O}(N^{2/3} \log(N))$ memory units, if assumptions (18) and (19) hold. □

In Theorem 4, we will perform the complexity analysis of the nearfield evaluation, i.e. lines 29 and 30 of Algorithm 4. In unbalanced trees there can be leaf clusters at coarse levels with large nearfields. If there were many of these, the complexity would not be linear. To exclude exceptional settings we make the additional assumption that the number of such leaf clusters is bounded, i.e., there exists a constant $c_{in} \in \mathbb{N}$ such that

$$\#\mathcal{L}_T^- \leq c_{in}, \quad \#\mathcal{L}_S^- \leq c_{in}, \tag{33}$$

where $\mathcal{L}_S^- := \mathcal{L}_S \setminus \mathcal{L}_S^+$ and

$$\begin{aligned}\mathcal{L}_S^+ := \{s \in \mathcal{L}_S \colon \#\{t : (t,s) \in \mathcal{L}_{T \times S}^-\} &\leq c_{nf} \text{ and} \\ \#\hat{t} &\leq c_m n_{max} \text{ for all } (t,s) \in \mathcal{L}_{T \times S}^-\},\end{aligned} \tag{34}$$

for some fixed parameters c_m and c_{nf}. It follows from (32) that for a leaf box $s \in \mathcal{T}_S^{(\ell)}$ the assumption $\#\{t : (t,s) \in \mathcal{L}_{T \times S}^-\} \leq c_{nf}$ holds true for sufficiently large constant c_{nf} if $\ell = \text{level}(t)$ is large enough.

Theorem 4. *Let assumptions (17) and (33) hold true. Then there exists a constant c_D depending only on c_{un}, c_{in}, c_m, c_{nf}, n_{max} and η_2 such that the number M_D of entries of all nearfield blocks $A|_{\hat{t} \times \hat{s}}$ in Algorithm 4 is bounded by*

$$M_D \leq c_D(N_T + N_S + 8^{p_{min}}). \tag{35}$$

If (18) holds, the corresponding storage complexity is $\mathcal{O}(N)$.

Proof. Each nearfield matrix block corresponds to an inadmissible block $(t,s) \in \mathcal{L}_{T \times S}^-$. For such a block there holds $t \in \mathcal{L}_T$ or $s \in \mathcal{L}_S$ by construction. We start counting entries of blocks corresponding to leaves in $\mathcal{L}_S$ by considering the sets $\mathcal{L}_S^+$ and $\mathcal{L}_S^-$.

For the number $M_{D,S}^-$ of nearfield matrix entries corresponding to blocks (t,s) with outlying leaves $s \in \mathcal{L}_S^-$ there holds

$$M_{D,S}^- = \sum_{s \in \mathcal{L}_S^-} \#\hat{s} \sum_{\{t:(t,s) \in \mathcal{L}_{T \times S}^-\}} \#\hat{t} \leq c_{in} n_{max} N_T. \tag{36}$$

Here we used that $\#\hat{s} \leq n_{max}$ holds for all leaf boxes, and that the nearfield $\mathcal{N}(s) = \{t : (t,s) \in \mathcal{L}_{T \times S}^-\}$ of s can contain at most all N_T points in P_T.

Next we estimate the number $M_{D,S}^+$ of nearfield matrix entries corresponding to blocks (t,s) with $s \in \mathcal{L}_S^+$. For fixed $s \in \mathcal{L}_S^+$ there exist at most c_{nf} such blocks (t,s) and the corresponding boxes t contain maximally $c_m n_{max}$ points by definition of $\mathcal{L}_S^+$ in (34). Furthermore, the level of a box s in an inadmissible block (t,s) can be at most p_{min} and $\mathcal{T}_S$ can have at most $8^{p_{min}}$ leaves at levels $\ell \leq p_{min}$. Hence, we get

$$M_{D,S}^+ = \sum_{s \in \mathcal{L}_S^+} \#\hat{s} \sum_{t \in \mathcal{N}(s)} \#\hat{t} \leq 8^{p_{min}} c_{nf} c_m n_{max}^2. \tag{37}$$

Analogous estimates as (36) and (37) hold true for nearfield matrices corresponding to leaves in $\mathcal{L}_T$. Adding up all these estimates leads to the bound in (35), with the constant $c_D = 2n_{max} \max\{c_{in}, c_{nf}c_m n_{max}\}$. If (18) holds, the storage complexity $\mathcal{O}(N)$ is an immediate consequence of (35).   □

The following theorem summarizes the results of this section.

Theorem 5. *Let assumptions (16)–(20) and (33) hold true. Then the complexity of Algorithm 4 is $\mathcal{O}(N \log(N))$. If (19) is replaced by (25) the complexity is reduced to $\mathcal{O}(N)$.*

4   Numerical Examples

In this section we want to test the method presented in Sect. 2 and to validate the theoretical results from Sect. 3. For this purpose we use a single core implementation of Algorithm 4 in C++ on a computer with 384 GiB RAM and 2 Intel Xeon Gold 5218 CPUs. To reduce the required memory we store only the non-directional parts $E_{t'}$ of the transfer matrices and each coupling matrix once, as described in Sect. 2.7. However, if the matrix is applied several times it can be beneficial to store also the directional interpolation matrices $L_{t,c}$ and nearfield matrix blocks.

For the tests we consider points distributed uniformly inside the cube $[-1, 1]^3$. For various values $k \geq 3$ we choose $\tilde{x}_n = (2n - 1)2^{-k} - 1$ in $[-1, 1]$ for all $n \in \{1, ..., 2^k\}$ and construct the set of points $P_T(k) = \{x_j\}_{j=1}^{N(k)}$ with $N(k) = 8^k$ as tensor products of these one-dimensional points. We choose $P_S(k) = P_T(k)$ and consider the matrix A as in (1) with the wave number $\kappa = 0.1 \cdot 2^k$ and the diagonal set to zero to eliminate the singularities. The approximation derived in Sect. 2 is applicable despite the change of the diagonal because it effects only parts of the matrix which are evaluated directly.

We construct a uniform box cluster tree $\mathcal{T}_T$ for the set P_T using Algorithm 1 with the initial box $T = [-1, 1]^3$ and the parameter $n_{max} = 512$. With this choice of parameters and points, $\mathcal{T}_T$ is a uniform octree with depth $p(\mathcal{T}_T) = k-3$, where every leaf contains exactly 512 points. We construct the sets of directions $D^{(\ell)}$ with Algorithm 3 and the largest high frequency level $\ell_{hf} = k - 4$ and finally we use Algorithm 2 to construct the block cluster tree $\mathcal{T}_{T \times T}$ with the parameter $\eta_2 = 5$. The parameters ℓ_{hf} and η_2 were chosen according to the parameter choice rule in [17, Sect. 3.1.4]. In particular, the choice $\eta_2 = 5$ minimizes the number of inadmissible blocks $b \in \mathcal{L}_{T \times T}^-$ at levels $\ell > \ell_{hf}$. Note that due to the uniformity of the tree $\mathcal{T}_T$ and the choice $P_S(k) = P_T(k)$ the block tree $\mathcal{T}_{T \times T}$ has depth $p(\mathcal{T}_T) = \ell_{hf} + 1$ and all inadmissible blocks are at level $\ell_{hf} + 1$.

The assumptions (16)–(20) are all satisfied for the considered examples for suitable constants c_{max}, c_{un}, c_{ad}, c_{geo}, and c_{hf} independent of the sets $P_T(k)$. Assumption (33) holds for $c_{in} = 0$, because all leafs in $\mathcal{L}_T$ are at level $k - 3$ and by the choice of η_2 there holds $\#\{s : (t, s) \in \mathcal{L}_{T \times T}^-\} \leq 27$ for all leaves $t \in \mathcal{L}_T$.

In the described setting we apply Algorithm 4 for the fast multiplication of the matrix A with a randomly constructed vector v. The interpolation degree

Table 1. Computation times and storage requirements for matrix-vector multiplications using Algorithm 4 for the matrix A corresponding to sets of points $P_T(k)$ for various values of k. Parameters: $\ell_{\mathrm{hf}} = k - 4$, $\eta_2 = 5$, interpolation degree $m = 4$.

k	N	κ	t_{tot}	t_{s}	t_{nf}	t_{ff}	nf [%]	N_{SC}	N_{C}	[GiB]
5	32768	3.2	7.31	0.34	6.94	0.03	24.41	316	3096	0.02
6	262144	6.4	76.25	1.20	74.29	0.76	4.06	1522	166320	0.10
7	2097152	12.8	702.72	3.71	688.14	10.87	0.58	4554	2640960	0.46
8	16777216	25.6	6060.16	15.24	5907.66	137.26	0.077	9824	33103296	3.09
9	134217728	51.2	50204.00	118.89	48576.20	1508.91	0.010	32036	344979432	24.2

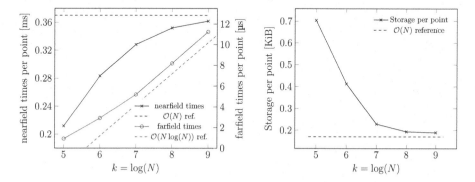

Fig. 1. Plots related to the computations of Table 1. Left image: Nearfield and farfield computation times per point with linear and quasi-linear reference curves in the different scales. Right image: Required storage per point with linear reference curve.

$m = 4$ is chosen, since it is reasonably high to yield a good approximation quality (e.g. relative error $2 \cdot 10^{-4}$ for $k = 6$) while it is low enough to make the approximations of all admissible blocks efficient.

The results of the computations for various sets of points $P_T(k)$ are given in Table 1 and Fig. 1. The total computational times t_{tot} are split into setup times t_{s}, times t_{nf} of the nearfield part, and computational times t_{ff} of the farfield part. In addition, the percentage of matrix entries in inadmissible blocks (nf), the numbers N_{SC} and N_{C} of stored and applied coupling matrices and the storage requirements ([GiB]) are given. A direct computation for $k = 7$ takes more than 32 hours. Thus the directional approximation is about 160 times faster. For larger examples the difference would be even more pronounced due to the quadratic complexity of the direct computation.

In Fig. 1, we plot computational times and memory consumption per point. As expected from our theoretical results of Sect. 3, we observe linear and almost linear behavior, respectively, for the nearfield and the farfield part of the computations, see the left plot in Fig. 1. As usual there is some preasymptotic behavior in such plots. The right plot in Fig. 1 shows the linear behavior of the memory requirements. Note that we store coupling matrices and transfer matrices only. In particular, we mention the low number N_{SC} of stored coupling matrices compared to the total number N_C of coupling matrices in Table 1.

Acknowledgment. This work was partially supported by the Austrian Science Fund (FWF): I 4033-N32.

References

1. Bebendorf, M., Kuske, C., Venn, R.: Wideband nested cross approximation for Helmholtz problems. Numer. Math. **130**(1), 1–34 (2015). https://doi.org/10.1007/s00211-014-0656-7
2. Bebendorf, M., Rjasanow, S.: Adaptive low-rank approximation of collocation matrices. Computing **70**(1), 1–24 (2003). https://doi.org/10.1007/s00607-002-1469-6
3. Börm, S.: Directional $\mathcal{H}^2$-matrix compression for high-frequency problems. Numer. Linear. Algebra. Appl. **24**(6), e2112 (2017). https://doi.org/10.1002/nla.2112
4. Börm, S., Börst, C.: Hybrid matrix compression for high-frequency problems. SIAM J. Matrix Anal. Appl. **41**(4), 1704–1725 (2020). https://doi.org/10.1137/19M124280X
5. Börm, S., Melenk, J.M.: Approximation of the high-frequency Helmholtz kernel by nested directional interpolation: error analysis. Numer. Math. **137**(1), 1–34 (2017). https://doi.org/10.1007/s00211-017-0873-y
6. Brandt, A.: Multilevel computations of integral transforms and particle interactions with oscillatory kernels. Comput. Phys. Commun. **65**(1), 24–38 (1991). https://doi.org/10.1016/0010-4655(91)90151-A
7. Cheng, H., et al.: A wideband fast multipole method for the Helmholtz equation in three dimensions. J. Comput. Phys. **216**, 300–325 (2006). https://doi.org/10.1016/j.jcp.2005.12.001
8. Darve, E., Havé, P.: Efficient fast multipole method for low-frequency scattering. J. Comput. Phys. **197**(1), 341–363 (2004). https://doi.org/10.1016/j.jcp.2003.12.002
9. Engquist, B., Ying, L.: Fast directional multilevel algorithms for oscillatory kernels. SIAM J. Sci. Comput **29**(4), 1710–1737 (2007). https://doi.org/10.1137/07068583X
10. Greengard, L., Rokhlin, V.: A fast algorithm for particle simulations. J. Comput. Phys. **73**(2), 325–348 (1987). https://doi.org/10.1016/0021-9991(87)90140-9
11. Hackbusch, W.: Hierarchical matrices: algorithms and analysis, SSCM, vol. 49. Springer, Heidelberg (2015). https://doi.org/10.1007/978-3-662-47324-5
12. Hu, B., Chew, W.C.: Fast inhomogeneous plane wave algorithm for scattering from objects above the multilayered medium. IEEE Trans. Geosci. Remote Sens. **39**(5), 1028–1038 (2001). https://doi.org/10.1109/36.921421
13. Messner, M., Schanz, M., Darve, E.: Fast directional multilevel summation for oscillatory kernels based on Chebyshev interpolation. J. Comput. Phys. **231**(4), 1175–1196 (2012). https://doi.org/10.1016/j.jcp.2011.09.027
14. Nishimura, N.: Fast multipole accelerated boundary integral equation methods. Appl. Mech. Rev. **55**(4), 299–324 (2002). https://doi.org/10.1115/1.1482087
15. Rjasanow, S., Steinbach, O.: The Fast Solution of Boundary Integral Equations. Springer-Verlag, Berlin, Heidelberg (2007). https://doi.org/10.1007/0-387-34042-4

16. Rokhlin, V.: Diagonal forms of translation operators for Helmholtz equation in three dimensions. Appl. Comput. Harmon. A. **1**, 82–93 (1993). https://doi.org/10.1006/acha.1993.1006

17. Watschinger, R.: A directional approximation of the Helmholtz kernel and its application to fast matrix-vector multiplications. Master's thesis, Graz University of Technology, Insitute of Applied Mathematics (2019). https://permalink.obvsg.at/AC15364438

Fast Large-Scale Boundary Element Algorithms

Steffen Börm[✉][iD]

Mathematisches Seminar, Christian-Albrechts-Universität zu Kiel,
24118 Kiel, Germany
boerm@math.uni-kiel.de
http://www.math.uni-kiel.de/scicom

Abstract. Boundary element methods (BEM) reduce a partial differential equation in a domain to an integral equation on the domain's boundary. They are particularly attractive for solving problems on unbounded domains, but handling the dense matrices corresponding to the integral operators requires efficient algorithms.

This article describes two approaches that allow us to solve boundary element equations on surface meshes consisting of several millions of triangles while preserving the optimal convergence rates of the Galerkin discretization.

Keywords: Boundary element method · Hierarchical matrices · Rank-structured matrices · Fast solvers

1   Introduction

We consider Laplace's equation

$$\Delta u(x) = 0 \qquad\qquad \text{for all } x \in \Omega, \qquad (1a)$$

where $\Omega \subseteq \mathbb{R}^3$ is a non-empty domain with a sufficiently smooth boundary $\partial\Omega$. If we add the boundary condition

$$u(x) = f(x) \qquad\qquad \text{for all } x \in \partial\Omega, \qquad (1b)$$

with a suitable function f, we obtain the *Dirichlet problem*. With the boundary condition

$$\frac{\partial u}{\partial n}(x) = f(x) \qquad\qquad \text{for all } x \in \partial\Omega, \qquad (1c)$$

where n denotes the outward-pointing unit normal vector for the domain Ω, we arrive at the *Neumann problem*. We can solve these problems by using *Green's representation formula* (cf., e.g., [18, Theorem 2.2.2]) given by

$$u(x) = \int_{\partial\Omega} g(x,y)\frac{\partial u}{\partial n}(y)\,dy - \int_{\partial\Omega} \frac{\partial g}{\partial n_y}(x,y)u(y)\,dy \qquad \text{for all } x \in \Omega, \qquad (2)$$

© Springer Nature Switzerland AG 2021
T. Kozubek et al. (Eds.): HPCSE 2019, LNCS 12456, pp. 60–79, 2021.
https://doi.org/10.1007/978-3-030-67077-1_4

where

$$g(x, y) = \begin{cases} \frac{1}{4\pi\|x-y\|} & \text{if } x \neq y, \\ 0 & \text{otherwise} \end{cases} \qquad \text{for all } x, y \in \mathbb{R}^3$$

denotes the free-space Green's function for the Laplace operator. If we know Dirichlet and Neumann boundary conditions, we can use (2) to compute the solution u in any point of the domain Ω.

For boundary values $x \in \partial\Omega$, Green's formula takes the form

$$\frac{1}{2}u(x) = \int_{\partial\Omega} g(x, y)\frac{\partial u}{\partial n}(y)\, dy - \int_{\partial\Omega} \frac{\partial g}{\partial n_y}(x, y)u(y)\, dy$$
$$\text{for almost all } x \in \partial\Omega, \tag{3}$$

at least in the distributional sense [27, eq. (3.92)], and this boundary integral equation can be used to solve the Dirichlet problem: we can solve the integral equation to obtain the Neumann values $\frac{\partial u}{\partial n}$ from the Dirichlet values and then use (2) to find the solution u in all of Ω.

In order to solve the Neumann problem, we take the normal derivative of (3) and arrive at the boundary integral equation

$$\frac{1}{2}\frac{\partial u}{\partial n}(x) = \frac{\partial}{\partial n_x}\int_{\partial\Omega} g(x, y)\frac{\partial u}{\partial n}(y)\, dy - \frac{\partial}{\partial n_x}\int_{\partial\Omega} \frac{\partial g}{\partial n_y}(x, y)u(y)\, dy$$
$$\text{for almost all } x \in \partial\Omega, \tag{4}$$

again in the distributional sense [27, eq. (3.92)], that can be used to find Dirichlet values matching given Neumann values, so we can follow the same approach as for the Dirichlet problem. The Dirichlet values are only determined up to a constant by the Neumann values, and factoring out the constants leads to unique solutions [27, Theorem 3.5.3].

In this paper, we will concentrate on solving the boundary integral equations (3) and (4) efficiently. We employ a Galerkin scheme using a finite-dimensional space V_h spanned by basis functions $(\psi_i)_{i \in \mathcal{I}}$ for the Neumann values and another finite-dimensional space U_h spanned by basis functions $(\varphi_j)_{j \in \mathcal{J}}$ for the Dirichlet values. The discretization turns the boundary integral operators into matrices, i.e., the matrix $G \in \mathbb{R}^{\mathcal{I} \times \mathcal{I}}$ corresponding to the *single-layer operator* given by

$$g_{ij} = \int_{\partial\Omega} \psi_i(x)\int_{\partial\Omega} g(x, y)\psi_j(y)\, dy\, dx \qquad \text{for all } i, j \in \mathcal{I}, \tag{5a}$$

the matrix $K \in \mathbb{R}^{\mathcal{I} \times \mathcal{J}}$ corresponding to the *double-layer operator* given by

$$k_{ij} = \int_{\partial\Omega} \psi_i(x)\int_{\partial\Omega} \frac{\partial g}{\partial n_y}(x, y)\varphi_j(y)\, dy\, dx \qquad \text{for all } i \in \mathcal{I}, \ j \in \mathcal{J}, \tag{5b}$$

and the matrix $W \in \mathbb{R}^{\mathcal{J} \times \mathcal{J}}$ corresponding to the *hypersingular operator* given by

$$w_{ij} = -\int_{\partial\Omega} \varphi_i(x)\frac{\partial}{\partial n_x}\int_{\partial\Omega} \frac{\partial g}{\partial n_y}(x, y)\varphi_j(y)\, dy\, dx \qquad \text{for all } i, j \in \mathcal{J}. \tag{5c}$$

In order to set up this last matrix, we use an alternative representation [27, Corollary 3.3.24] based on the single-layer operator.

Together with the mixed mass matrix $M \in \mathbb{R}^{\mathcal{I} \times \mathcal{J}}$ given by

$$m_{ij} = \int_{\partial\Omega} \psi_i(x)\varphi_j(y) \qquad \text{for all } i \in \mathcal{I}, \ j \in \mathcal{J},$$

we obtain the linear system

$$Gx = \left(\frac{1}{2}M + K\right)b \tag{6}$$

for the Dirichlet-to-Neumann problem (3), where $x \in \mathbb{R}^{\mathcal{I}}$ contains the coefficients of the Neumann values and $b \in \mathbb{R}^{\mathcal{J}}$ those of the given Dirichlet values, and the system

$$Wx = \left(\frac{1}{2}M^* - K^*\right)b \tag{7}$$

for the Neumann-to-Dirichlet problem (4), where $x \in \mathbb{R}^{\mathcal{J}}$ now contains the coefficients of the Dirichlet values and $b \in \mathbb{R}^{\mathcal{I}}$ those of the given Neumann values. M^* and K^* denote the transposed matrices of M and K, respectively.

Considered from the point of view of numerical mathematics, these linear systems pose three challenges: we have to evaluate singular integrals in order to compute diagonal and near-diagonal entries, the resulting matrices are typically dense and large, and the condition number grows with the matrix dimension, which leads to a deteriorating performance of standard Krylov solvers. The first challenge can be met by using suitable quadrature schemes like the Sauter-Schwab-Erichsen technique [12,26,27]. For the third challenge, various preconditioning algorithms have been proposed [23,28], among which we choose $\mathcal{H}$-matrix coarsening and factorization [2,14].

This leaves us with the second challenge, i.e., the efficient representation of the dense matrices G, K, and W. This task is usually tackled by employing a data-sparse approximation, i.e., by finding a sufficiently accurate approximation of the matrices that requires only a small amount of data. One possibility to construct such an approximation is to use wavelet basis functions and neglect small matrix entries in order to obtain a sparse matrix [10,22,24,29]. Since the construction of suitable wavelet spaces on general surfaces is complicated [11], we will not focus on this approach.

Instead, we consider low-rank approximation techniques that directly provide an approximation of the matrices for standard finite element basis functions. These techniques broadly fall into three categories: *analytic* methods approximate the kernel function g locally by sums of tensor products, and discretization of these sums leads to low-rank approximations of matrix blocks. The most prominent analytic methods are the fast multipole expansion [16,17,25] and interpolation [8,13]. *Algebraic* methods, on the other hand, directly approximate the matrix blocks, e.g., by computing a rank-revealing factorization like an *adaptive cross approximation* [1,3,4]. The convergence and robustness of analytic methods can be proven rigorously, but they frequently require a larger than necessary amount of storage. Algebraic methods reach close to optimal compression, but involve a heuristic pivoting strategy that may fail in certain cases [9, Example 2.2]. *Hybrid* methods combine analytic and algebraic techniques in order to obtain the advantages of both without the respective disadvantages. In this paper, we will focus on the *hybrid cross approximation* [9] and the *Green cross approximation* [7] that both combine an analytic approximation with an algebraic improvement in order to obtain fast and reliable algorithms.

2 $\mathcal{H}^2$-Matrices

Both approximation schemes in this paper's focus lead to $\mathcal{H}^2$-*matrices* [6,21], a special case of *hierarchical matrices* [15,19]. A given matrix $G \in \mathbb{R}^{\mathcal{I} \times \mathcal{J}}$ with general finite row and column index sets $\mathcal{I}$ and $\mathcal{J}$ is split into submatrices that are approximated by factorized low-rank representations.

The submatrices are constructed hierarchically in order to make the matrix accessible for elegant and efficient recursive algorithms. The first step is to split the index sets into a tree structure of subsets.

Definition 1 (Cluster tree). *Let $\mathcal{I}$ be a finite non-empty index set. A tree $\mathcal{T}$ is called a* cluster tree *for $\mathcal{I}$ if the following conditions hold:*

- *Every node of the tree is a subset of $\mathcal{I}$.*
- *The root is $\mathcal{I}$.*
- *If a node has children, it is the union of these children:*

$$t = \bigcup_{t' \in \mathrm{chil}(t)} t' \qquad \text{for all } t \in \mathcal{T} \text{ with } \mathrm{chil}(t) \neq \emptyset.$$

- *The children of a node $t \in \mathcal{T}$ are disjoint:*

$$t_1 \cap t_2 \neq \emptyset \Rightarrow t_1 = t_2 \qquad \text{for all } t \in \mathcal{T}, \ t_1, t_2 \in \mathrm{chil}(t).$$

Nodes of a cluster tree are called clusters.

Cluster trees can be constructed by recursively splitting index sets, e.g., ensuring that "geometrically close" indices are contained in the same cluster [20, Section 5.4]. We assume that cluster trees $\mathcal{T}_{\mathcal{I}}$ and $\mathcal{T}_{\mathcal{J}}$ for the index sets $\mathcal{I}$ and $\mathcal{J}$ are given.

Using the cluster trees, the index set $\mathcal{I} \times \mathcal{J}$ corresponding to the matrix G can now be split into a tree structure.

Definition 2 (Block tree). *A tree $\mathcal{T}$ is called a* block tree *for the row index set $\mathcal{I}$ and the column index set $\mathcal{J}$ if the following conditions hold:*

- *Every node of the tree is a subset $t \times s \subseteq \mathcal{I} \times \mathcal{J}$ with $t \in \mathcal{T}_{\mathcal{I}}$ and $s \in \mathcal{T}_{\mathcal{J}}$.*
- *The root is $\mathcal{I} \times \mathcal{J}$.*

– *If a node has children, the children are given as follows:*

$$\mathrm{chil}(t \times s) = \begin{cases} \{t \times s' \ : \ s' \in \mathrm{chil}(s)\} & \textit{if } \mathrm{chil}(t) = \emptyset, \\ \{t' \times s \ : \ t' \in \mathrm{chil}(t)\} & \textit{if } \mathrm{chil}(s) = \emptyset, \\ \{t' \times s' \ : \ t' \in \mathrm{chil}(t), \ s' \in \mathrm{chil}(s)\} & \textit{otherwise} \end{cases}$$

for all $t \times s \in \mathcal{T}$.

Nodes of a block tree are called blocks.

Block trees are usually constructed recursively using an *admissibility condition* matching the intended approximation scheme: we start with the root $\mathcal{I} \times \mathcal{J}$ and recursively subdivide blocks. Once the admissibility condition indicates for a block $t \times s$ that we can approximate the submatrix $G|_{t \times s}$, we stop subdividing and call the block $t \times s$ a *farfield block*. We also stop if t and s have no children, then $t \times s$ is a *nearfield block*. If we ensure that leaf clusters contain only a small number of indices, nearfield blocks are small and we can afford to store them without compression. The key to the efficiency of hierarchical matrices is the data-sparse representation of the farfield blocks.

We assume that a block tree $\mathcal{T}_{\mathcal{I} \times \mathcal{J}}$ is given and that its farfield leaves are collected in a set $\mathcal{L}^+_{\mathcal{I} \times \mathcal{J}}$, while the nearfield leaves are in a set $\mathcal{L}^-_{\mathcal{I} \times \mathcal{J}}$.

For a hierarchical matrix [15, 19], we simply assume that farfield blocks have low rank $k \in \mathbb{N}$ and can therefore be stored efficiently in factorized form $G|_{t \times s} \approx A_{ts} B^*_{ts}$ with $A_{ts} \in \mathbb{R}^{t \times k}$, $B_{ts} \in \mathbb{R}^{s \times k}$. Here we use $\mathbb{R}^{t \times k}$ and $\mathbb{R}^{s \times k}$ as abbreviations for $\mathbb{R}^{t \times [1:k]}$ and $\mathbb{R}^{s \times [1:k]}$.

The more efficient $\mathcal{H}^2$-matrices [6, 21] use a different factorized representation closely related to fast multipole methods [25]: each cluster is associated with a low-dimensional subspace, e.g., a space spanned by polynomials or multipole functions, and the range of a matrix block $G|_{t \times s}$ has to be contained in the subspace for the row cluster t, while the range of the adjoint block $G|^*_{t \times s}$ has to be contained in the subspace for the column cluster s. This property is expressed by the equation (9) below.

Definition 3 (Cluster basis). *Let $k \in \mathbb{N}$. A family $V = (V_t)_{t \in \mathcal{T}_{\mathcal{I}}}$ of matrices is called a* cluster basis *for $\mathcal{T}_{\mathcal{I}}$ with (maximal) rank k if the following conditions hold:*

– *We have $V_t \in \mathbb{R}^{t \times k}$ for all $t \in \mathcal{T}_{\mathcal{I}}$.*
– *For all $t \in \mathcal{T}_{\mathcal{I}}$ with $\mathrm{chil}(t) \neq \emptyset$, there are transfer matrices $E_{t'} \in \mathbb{R}^{k \times k}$ for all $t' \in \mathrm{chil}(t)$ such that*

$$V_t|_{t' \times k} = V_{t'} E_{t'} \qquad \textit{for all } t' \in \mathrm{chil}(t). \qquad (8)$$

An important property of cluster bases is that they can be stored efficiently: under standard assumptions, only $\mathcal{O}(nk)$ units of storage are requires, where $n = |\mathcal{I}|$ is the cardinality of the index set $\mathcal{I}$.

Definition 4. ($\mathcal{H}^2$-matrix). *Let $V = (V_t)_{t \in \mathcal{T}_\mathcal{I}}$ and $W = (W_s)_{s \in \mathcal{T}_\mathcal{J}}$ be cluster bases for $\mathcal{T}_\mathcal{I}$ and $\mathcal{T}_\mathcal{J}$, respectively. A matrix $G \in \mathbb{R}^{\mathcal{I} \times \mathcal{J}}$ is called an $\mathcal{H}^2$-matrix with row basis $(V_t)_{t \in \mathcal{T}_\mathcal{I}}$ and column basis $(W_s)_{s \in \mathcal{T}_\mathcal{J}}$ if for every admissible leaf $t \times s \in \mathcal{L}^+_{\mathcal{I} \times \mathcal{J}}$ there is a coupling matrix $S_{ts} \in \mathbb{R}^{k \times k}$ such that*

$$G|_{t \times s} = V_t S_{ts} W_s^*. \tag{9}$$

Under standard assumptions, we can represent an $\mathcal{H}^2$-matrix by $\mathcal{O}(nk + mk)$ coefficients, where $n = |\mathcal{I}|$ and $m = |\mathcal{J}|$ are the cardinalities of the index sets. The matrix-vector multiplication $x \mapsto Gx$ can be performed in $\mathcal{O}(nk + mk)$ operations for $\mathcal{H}^2$-matrices, and there are a number of other important operations that also only have linear complexity with respect to n and m, cf. [6].

3   Hybrid Cross Approximation

In order to construct an $\mathcal{H}^2$-matrix approximation of the matrices V, K, and W required for the boundary element method, we first consider the *hybrid cross approximation* (HCA) method [9] originally developed for hierarchical matrices.

We associate each cluster $t \in \mathcal{T}_\mathcal{I}$ with an axis-parallel *bounding box* $B_t \subseteq \mathbb{R}^3$ such that the supports of all basis functions associated with indices in t are contained in B_t. In order to ensure that the kernel function g is sufficiently smooth for a polynomial approximation, we introduce the admissibility condition

$$\max\{\mathrm{diam}(B_t), \mathrm{diam}(B_s)\} \leq 2\eta \, \mathrm{dist}(B_t, B_s), \tag{10}$$

where $\mathrm{diam}(B_t)$ and $\mathrm{diam}(B_s)$ denote the Euclidean diameters of the bounding boxes B_t and B_s, while $\mathrm{dist}(B_t, B_s)$ denotes their Euclidean distance. η is a parameter that controls the storage complexity and the accuracy of the approximation. In our experiments, the choice $\eta = 1$ leads to reasonable results.

If two clusters $t \in \mathcal{T}_\mathcal{I}$, $s \in \mathcal{T}_\mathcal{J}$ satisfy this condition, the restriction $g|_{B_t \times B_s}$ can be approximated by polynomials. We use m-th order tensor Chebyshev interpolation and denote the interpolation points for B_t and B_s by $(\xi_{t,\nu})_{\nu=1}^k$ and $(\xi_{s,\mu})_{\mu=1}^k$, where $k = m^3$. The corresponding Lagrange polynomials are denoted by $(\mathcal{L}_{t,\nu})_{\nu=1}^k$ and $(\mathcal{L}_{s,\mu})_{\mu=1}^k$, and the tensor interpolation polynomial is given by

$$\tilde{g}_{\mathrm{int}}(x, y) = \sum_{\nu=1}^k \sum_{\mu=1}^k \mathcal{L}_{t,\nu}(x) g(\xi_{t,\nu}, \xi_{s,\mu}) \mathcal{L}_{s,\mu}(y) \qquad \text{for all } x \in B_t, \ y \in B_s.$$

It is possible to prove

$$\|g - \tilde{g}_{\mathrm{int}}\|_{\infty, B_t \times B_s} \lesssim \frac{q^m}{\mathrm{diam}(B_t)^{1/2} \, \mathrm{diam}(B_s)^{1/2}}$$

$$\text{for all } m \in \mathbb{N} \text{ and all } t \in \mathcal{T}_\mathcal{I}, s \in \mathcal{T}_\mathcal{J} \text{ satisfying (10)}.$$

The rate q of convergence depends only on the parameter η, cf. [6, Chapter 4].

Unfortunately, the rank k of this approximation is too large: potential theory suggests that a rank $k \sim m^2$ should be sufficient for an m-th order approximation.

We reduce the rank by combining the interpolation with an algebraic procedure, in this case *adaptive cross approximation* [1,4]: we introduce the matrix $S \in \mathbb{R}^{k \times k}$ by

$$s_{\nu\mu} = g(\xi_{t,\nu}, \xi_{s,\mu}) \qquad \text{for all } \nu, \mu \in [1:k]$$

and use a rank-revealing pivoted LU factorization to obtain an approximation of the form

$$S \approx S|_{[1:k] \times \sigma} C S|_{\tau \times [1:k]},$$

where $C := (S|_{\tau \times \sigma})^{-1}$ and $\tau, \sigma \subseteq [1:k]$ denote the first $\tilde{k} \leq k$ row and column pivots, respectively. Due to the properties of the kernel function g, the rank $\tilde{k}$ can be expected to be significantly smaller than k.

Given the pivot sets τ and σ, we can now "take back" the interpolation:

$$g(x,y) \approx \sum_{\nu=1}^{k} \sum_{\mu=1}^{k} \mathcal{L}_{t,\nu}(x) s_{\nu\mu} \mathcal{L}_{s,\mu}(y)$$

$$\approx \sum_{\nu=1}^{k} \sum_{\mu=1}^{k} \mathcal{L}_{t,\nu}(x) (S|_{[1:k] \times \sigma} C S|_{\tau \times [1:k]})_{\nu\mu} \mathcal{L}_{s,\mu}(y)$$

$$= \sum_{\nu=1}^{k} \sum_{\mu=1}^{k} \sum_{\lambda \in \tau} \sum_{\kappa \in \sigma} \mathcal{L}_{t,\nu}(x) s_{\nu\kappa} c_{\kappa\lambda} s_{\lambda\mu} \mathcal{L}_{s,\mu}(y)$$

$$= \sum_{\lambda \in \tau} \sum_{\kappa \in \sigma} \underbrace{\sum_{\nu=1}^{k} \mathcal{L}_{t,\nu}(x) g(\xi_{t,\nu}, \xi_{s,\kappa})}_{\approx g(x, \xi_{s,\kappa})} c_{\kappa\lambda} \underbrace{\sum_{\mu=1}^{k} g(\xi_{t,\lambda}, \xi_{s,\mu}) \mathcal{L}_{s,\mu}(y)}_{\approx g(\xi_{t,\lambda}, y)}$$

$$\approx \sum_{\lambda \in \tau} \sum_{\kappa \in \sigma} g(x, \xi_{s,\kappa}) c_{\kappa\lambda} g(\xi_{t,\lambda}, y). =: \tilde{g}_{\text{hca}}(x,y)$$

By controlling the interpolation order and the accuracy of the cross approximation, we can ensure that the approximation error $\|g - \tilde{g}_{\text{hca}}\|_{\infty, B_t \times B_s}$ is below any given tolerance [9].

In order to obtain an approximation of the submatrix $G|_{t\times s}$, we replace g by $\tilde{g}_{hca}$ in (5a) to find

$$
\begin{aligned}
g_{ij} &= \int_{\partial\Omega} \psi_i(x) \int_{\partial\Omega} g(x,y)\psi_j(y)\,dy\,dx \\
&\approx \int_{\partial\Omega} \psi_i(x) \int_{\partial\Omega} \tilde{g}_{hca}(x,y)\psi_j(y)\,dy\,dx \\
&= \sum_{\lambda\in\tau}\sum_{\kappa\in\sigma} \underbrace{\int_{\partial\Omega} \psi_i(x)g(x,\xi_{s,\kappa})\,c_{\lambda\kappa}}_{=:a_{i\kappa}} \underbrace{\int_{\partial\Omega} \psi_j(y)g(\xi_{t,\lambda},y)}_{=:b_{j\lambda}} \\
&= (ACB^*)_{ij} \qquad \text{for all } i\in t,\ j\in s
\end{aligned}
\tag{11}
$$

with matrices $A \in \mathbb{R}^{t\times\sigma}$ and $B \in \mathbb{R}^{s\times\tau}$. Due to $|\tau| = |\sigma| = \tilde{k} \le k$, the matrix ACB^* is an improved low-rank approximation of the submatrix $G|_{t\times s}$.

This procedure alone only yields a hierarchical matrix, not the desired $\mathcal{H}^2$-matrix. In order to reduce the storage requirements, we apply *hierarchical compression* [5]: the hybrid cross approximation technique already yields low-rank approximations of individual blocks, but these approximations do not share common row or column cluster bases. The hierarchical compression algorithm recursively merges independent submatrices into larger $\mathcal{H}^2$-matrices until the entire matrix is in the required form.

The parallelization of this algorithm is fairly straightforward: we can set up the matrices corresponding to the leaves of the block tree in parallel, and the merge operations for submatrices can also be performed in parallel once their children are available. In order to handle the dependencies between children and their parents, both clusters and blocks, a task-based programming model is particularly useful for this algorithm.

4 Green Cross Approximation

While hybrid cross approximation requires a subsequent compression step to obtain an $\mathcal{H}^2$-matrix, we now consider a related technique [7] that *directly* yields an $\mathcal{H}^2$-matrix approximation of the matrix G.

We once more consider an admissible pair $t \in \mathcal{T}_{\mathcal{I}}$, $s \in \mathcal{T}_{\mathcal{J}}$ of clusters with bounding boxes $B_t, B_s \subseteq \mathbb{R}^3$. We construct an auxiliary axis-parallel box $\omega_t \subseteq \mathbb{R}^3$ such that

$$
B_t \subseteq \omega_t, \qquad \text{diam}(\omega_t) \lesssim \text{dist}(B_t, \partial\omega_t), \qquad \text{diam}(\omega_t) \lesssim \text{dist}(B_s, \omega_t),
\tag{12}
$$

cf. Fig. 1 for an illustration. Let $y \in B_s$. The third assumption in (12) implies $y \notin \omega_t$, therefore the function $u(x) = g(x,y)$ is harmonic in ω_t. This property allows us to apply Green's equation (2) to the domain ω_t to obtain

$$
g(x,y) = \int_{\partial\omega_t} g(x,z)\frac{\partial g}{\partial n_z}(z,y)\,dz - \int_{\partial\omega_t} \frac{\partial g}{\partial n_z}(x,z)g(z,y)\,dz.
$$

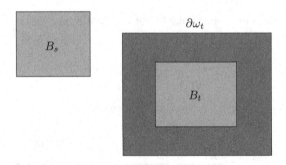

Fig. 1. Possible choice of the auxiliary bounding box ω_t corresponding to B_t and B_s

We observe that the variables x and y are separated in both integrands. Due to the second and third assumptions in (12), the integrands are smooth for $x \in B_t$ and $y \in B_s$, and we can approximate the integrals by a quadrature rule, e.g., a composite Gauss rule, with weights $(w_\nu)_{\nu=1}^k$ and quadrature points $(z_\nu)_{\nu=1}^k$ on the boundary $\partial\omega_t$ in order to get

$$g(x, y) \approx \sum_{\nu=1}^k w_\nu g(x, z_\nu) \frac{\partial g}{\partial n_z}(z_\nu, y) - \sum_{\nu=1}^k w_\nu \frac{\partial g}{\partial n_z}(x, z_\nu) g(z_\nu, y) =: \tilde{g}_{\mathrm{grn}}(x, y)$$

for all $x \in B_t$ and all $y \in B_s$. The function $\tilde{g}_{\mathrm{grn}}$ is again a sum of tensor products, and, as in the case of the hybrid cross approximation, its discretization gives rise to a low-rank approximation of the submatrix $G|_{t \times s}$, since we have

$$
\begin{aligned}
g_{ij} &= \int_{\partial\Omega} \psi_i(x) \int_{\partial\Omega} g(x, y) \psi_j(y)\, dy\, dx \\
&\approx \int_{\partial\Omega} \psi_i(x) \int_{\partial\Omega} \tilde{g}_{\mathrm{grn}}(x, y) \psi_j(y)\, dy\, dx \\
&= \sum_{\nu=1}^k \underbrace{w_\nu^{1/2} \int_{\partial\Omega} \psi_i(x) g(x, z_\nu)\, dx}_{=:a_{i\nu}} \underbrace{w_\nu^{1/2} \int_{\partial\Omega} \psi_j(y) \frac{\partial g}{\partial n_z}(z_\nu, y)\, dy}_{=:b_{j\nu}} \\
&\quad - \sum_{\nu=1}^k \underbrace{w_\nu^{1/2} \int_{\partial\Omega} \psi_i(x) \frac{\partial g}{\partial n_z}(x, z_\nu)\, dx}_{=:c_{i\nu}} \underbrace{w_\nu^{1/2} \int_{\partial\Omega} \psi_j(x) g(z_\nu, y)\, dy}_{=:d_{j\nu}} \\
&= (AB^* - CD^*)_{ij} \qquad \text{for all } i \in t,\ j \in s
\end{aligned}
$$

with $A, C \in \mathbb{R}^{t \times k}$ and $B, D \in \mathbb{R}^{s \times k}$, therefore

$$G|_{t \times s} \approx \begin{pmatrix} A & C \end{pmatrix} \begin{pmatrix} B^* \\ -D^* \end{pmatrix}.$$

Unfortunately, the rank of this approximation is quite high, higher than, e.g., for the hybrid cross approximation. Once again, we can use an algebraic technique, i.e., the adaptive cross approximation, to improve the analytically-motivated initial approximation $\tilde{g}_{grn}$: we define

$$L := \begin{pmatrix} A & C \end{pmatrix} \in \mathbb{R}^{t \times (2k)}$$

and perform an adaptive cross approximation to find index sets $\tau \subseteq t$ and $\sigma \subseteq [1{:}2k]$ of cardinality $\tilde{k} \leq 2k$ such that

$$L \approx L|_{t \times \sigma}(L|_{\tau \times \sigma})^{-1}L|_{\tau \times [1:2k]}.$$

By applying this approximation and "taking back" the quadrature approximation in the last step, we arrive at

$$G|_{t \times s} \approx L \begin{pmatrix} B^* \\ D^* \end{pmatrix} \approx L|_{t \times \sigma}(L|_{\tau \times \sigma})^{-1}L|_{\tau \times [1:2k]} \begin{pmatrix} B^* \\ D^* \end{pmatrix} \approx L|_{t \times \sigma}(L|_{\tau \times \sigma})^{-1}G|_{\tau \times s}.$$

We define $\hat{t} := \tau$ and $V_t := L|_{t \times \sigma}(L|_{\tau \times \sigma})^{-1} \in \mathbb{R}^{t \times \hat{t}}$ and obtain

$$G|_{t \times s} \approx V_t G|_{\hat{t} \times s}. \tag{13}$$

This is a rank-$\tilde{k}$ approximation of the matrix block, and experiments indicate that $\tilde{k}$ is frequently far smaller than $2k$. The equation (13) can be interpreted as "algebraic interpolation": we (approximately) recover all entries of the matrix $G|_{t \times s}$ from a few rows $G|_{\hat{t} \times s}$, where the indices in $\hat{t}$ play the role of interpolation points and the columns of V_t the role of Lagrange polynomials. We call this approach *Green cross approximation* (GCA).

Concerning our goal of finding an $\mathcal{H}^2$-matrix, we observe that V_t and $\hat{t}$ depend only on t and ω_t, but not on the cluster s, therefore V_t it is a good candidate for a cluster basis.

Applying the same procedure to the cluster s instead of t, we obtain $\hat{s} \subseteq s$ and $W_s \in \mathbb{R}^{s \times \hat{s}}$ such that

$$G|_{t \times s} \approx G|_{t \times \hat{s}}W_s^*,$$

and combining both approximations yields

$$G|_{t \times s} \approx V_t G|_{\hat{t} \times s} \approx V_t G|_{\hat{t} \times \hat{s}}W_s^*,$$

the required representation (9) for an $\mathcal{H}^2$-matrix.

In order to make $V = (V_t)_{t \in \mathcal{T}_\mathcal{I}}$ and $W = (W_s)_{s \in \mathcal{T}_\mathcal{J}}$ proper cluster bases, we have to ensure the nesting property (8). We can achieve this goal by slightly modifying our construction: we assume that $t \in \mathcal{T}_\mathcal{I}$ has two children $t_1, t_2 \in$ chil(t) and that the sets $\hat{t}_1 \subseteq t_1$ and $\hat{t}_2 \subseteq t_2$ and the matrices V_{t_1} and V_{t_2} have already been computed. We have

$$G|_{t \times s} = \begin{pmatrix} G|_{t_1 \times s} \\ G|_{t_2 \times s} \end{pmatrix} \approx \begin{pmatrix} V_{t_1} G|_{\hat{t}_1 \times s} \\ V_{t_2} G|_{\hat{t}_2 \times s} \end{pmatrix} = \begin{pmatrix} V_{t_1} & \\ & V_{t_2} \end{pmatrix} G|_{(\hat{t}_1 \cup \hat{t}_2) \times s}.$$

We apply the cross approximation to $G|_{(\hat{t}_1 \cup \hat{t}_2) \times s}$ instead of $G|_{t \times s}$ and obtain $\hat{t} \subseteq \hat{t}_1 \cup \hat{t}_2$ and $\widehat{V}_t \in \mathbb{R}^{(\hat{t}_1 \cup \hat{t}_2) \times \hat{t}}$ such that

$$G|_{(\hat{t}_1 \cup \hat{t}_2) \times s} \approx \widehat{V}_t G|_{\hat{t} \times s}$$

and therefore

$$G|_{t \times s} \approx \begin{pmatrix} V_{t_1} & \\ & V_{t_2} \end{pmatrix} G|_{(\hat{t}_1 \cup \hat{t}_2) \times s} \approx \begin{pmatrix} V_{t_1} & \\ & V_{t_2} \end{pmatrix} \widehat{V}_t G|_{\hat{t} \times s},$$

so defining

$$V_t := \begin{pmatrix} V_{t_1} & \\ & V_{t_2} \end{pmatrix} \widehat{V}_t$$

ensures (8) if we let

$$\begin{pmatrix} E_{t_1} \\ E_{t_2} \end{pmatrix} := \widehat{V}_t.$$

This modification not only ensures that $V = (V_t)_{t \in \mathcal{T}_{\mathcal{I}}}$ is a proper cluster basis, it also reduces the computational work required for the cross approximation, since only the submatrices $G|_{(\hat{t}_1 \cup \hat{t}_2) \times s}$ have to be considered, and these submatrices are significantly smaller than $G|_{t \times s}$.

The parallelization of this algorithm is straightforward: we can compute index sets and matrices for all leaves of the cluster tree in parallel. Once this task has been completed, we can treat all clusters whose children are leaves in parallel. Next are all clusters whose children have already been treated. Repeating this procedure until we reach the root of the tree yields a simple and efficient parallel version of the algorithm. Our implementation uses simple parallel for loops to iterate through the children of clusters and blocks prior to setting up their parents.

5   Numerical Experiments

Now that we have two compression algorithms at our disposal, we have to investigate how well they perform in practice. While we can prove for both algorithms that they can reach any given accuracy (disregarding rounding errors), we have to see which accuracies are necessary in order to preserve the theoretical convergence rates of the Galerkin discretization.

Until now, we have only seen HCA and GCA applied to the matrix G corresponding to the single-layer operator. For the other two operators, we simply take the appropriate derivatives of $\tilde{g}_{\text{hca}}$ and $\tilde{g}_{\text{grn}}$ and use them as approximations of the kernel functions.

Given a boundary element mesh, we choose discontinuous piecewise constant basis functions for the Neumann values and continuous piecewise linear basis functions for the Dirichlet values. For a meshwidth of $h \in \mathbb{R}_{>0}$, we expect theoretical convergence rates of $\mathcal{O}(h)$ for the Neumann values in the L^2 norm,

$\mathcal{O}(h^{3/2})$ for the Neumann values in the $H^{-1/2}$ norm and the Dirichlet values in the $H^{1/2}$ norm, and $\mathcal{O}(h^2)$ for the Dirichlet values in the L^2 norm.

In a first experiment, we approximate the unit sphere $\{x \in \mathbb{R}^3 : x_1^2 + x_2^2 + x_3^2 = 1\}$ by a sequence of triangular meshes constructed by splitting the eight sides of a double pyramid $\{x \in \mathbb{R}^3 : |x_1| + |x_2| + |x_3| = 1\}$ regularly into triangles and then projecting all vertices to the unit sphere. Since we expect the condition number of the matrices to be in $\mathcal{O}(h^{-1})$ and want to preserve the theoretical convergence rate of $\mathcal{O}(h^2)$, we aim for an accuracy of $\mathcal{O}(h^3)$ for the matrix approximation. For the sake of simplicity, we use a slightly higher accuracy: if h is halved, we reduce the error tolerance by a factor of 10 instead of just 8.

Nearfield matrix entries are computed by Sauter-Schwab-Erichsen quadrature [27], and we have to increase the order of the nearfield quadrature occasionally to ensure the desired rate of convergence.

Table 1. Parameters chosen for the unit sphere

n	q_{near}	r_{leaf}	m	ϵ_{aca}	ϵ_{comp}	ϵ_{slv}	ϵ_{prc}
8 192	4	25	5	1_{-5}	1_{-5}	1_{-6}	1_{-2}
18 432	4	36	6	3_{-6}	3_{-6}	3_{-7}	1_{-2}
32 768	4	36	6	1_{-6}	1_{-6}	1_{-7}	5_{-3}
73 728	5	49	7	3_{-7}	3_{-7}	3_{-8}	5_{-3}
131 072	5	49	7	1_{-7}	1_{-7}	1_{-8}	2_{-3}
294 912	5	64	8	3_{-8}	3_{-8}	3_{-9}	2_{-3}
524 288	5	64	8	1_{-8}	1_{-8}	1_{-9}	1_{-3}
1 179 648	6	81	9	3_{-9}	3_{-9}	3_{-10}	1_{-3}
2 097 152	6	81	9	1_{-9}	1_{-9}	1_{-10}	5_{-4}
4 718 592	6	100	10	3_{-10}	3_{-10}	3_{-11}	5_{-4}
8 388 608	6	100	10	1_{-10}	1_{-10}	1_{-11}	2_{-4}

The parameters used for this experiment are summarized in Table 1, where n denotes the number of triangles, q_{near} the nearfield quadrature order, r_{leaf} the resolution of the cluster tree, i.e., the maximal size of leaf clusters, m the order of interpolation for HCA and the order of quadrature for GCA, ϵ_{aca} the relative accuracy for the adaptive cross approximation, ϵ_{comp} the accuracy for the hierarchical compression used for HCA, ϵ_{slv} the relative accuracy of the Krylov solver, and ϵ_{prc} the relative accuracy of the preconditioner constructed by coarsening [14] and $\mathcal{H}$-Cholesky decomposition [20]. We use $\eta = 1$ for the admissibility parameter and construct the boxes ω_t for the Green quadrature to ensure $\mathrm{dist}(\partial\omega_t, B_t) = \delta_t$, with $\delta_t = \max\{b_1 - a_1, b_2 - a_2, b_3 - a_3\}$, where $B_t = [a_1, b_1] \times [a_2, b_2] \times [a_3, b_3]$. This choice is not strictly covered by the theory in [7], but works very well in practice. The nearfield order was only increased if the convergence was compromised. ϵ_{aca}, ϵ_{comp}, and ϵ_{slv} where chosen in the expectation that an accuracy of $\mathcal{O}(h^3)$ would be required in order to keep up with the discretization error. ϵ_{prc} was chosen in the expectation that an accuracy of $\mathcal{O}(h)$ would be necessary to keep up with the growth of the condition number of the linear system.

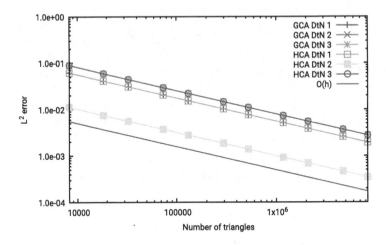

Fig. 2. L^2 error for the Dirichlet-to-Neumann problem

Figure 2 shows the L^2-norm error for the approximation of the Neumann data computed via the boundary integral equation (3). We use the functions $u_1(x) = x_1^2 - x_3^2$, $u_2(x) = g(x, y_1)$, and $u_3(x) = g(x, y_2)$ with $y_1 = (1.2, 1.2, 1.2)$ and $y_2 = (1.0, 0.25, 1.0)$ as test cases. We can see that the optimal convergence rate of $\mathcal{O}(h)$ is preserved despite the matrix compression and nearfield quadrature.

Figure 3 shows the L^2-norm error for the approximation of the Dirichlet data computed via the boundary integral equation (4). Also in this case, the optimal convergence rate of $\mathcal{O}(h^2)$ is preserved.

Since computing the exact $H^{-1/2}$-norm error is complicated, we rely on the $H^{-1/2}$-ellipticity of the single-layer operator: we compute the L^2-projection of the Neumann values into the discrete space, which is expected to converge at a rate of $\mathcal{O}(h^{3/2})$ to the exact solution, and then compare it to the Galerkin solution using the energy product corresponding to the matrix G. Figure 4 indicates that the discrete $H^{-1/2}$-norm error even converges at a rate of $\mathcal{O}(h^2)$, therefore the $\mathcal{O}(h^{3/2})$ convergence compared to the continuous solution is also preserved.

Now that we have established that the compression algorithms do not hurt the optimal convergence rates of the Galerkin discretization, we can consider the corresponding complexity. The runtimes have been measured on a system with two Intel® Xeon® Platinum 8160 processors, each with 24 cores and running at a base clock of 2.1 GHz. The implementation is based on the open-source H2Lib software package, cf. http://www.h2lib.org.

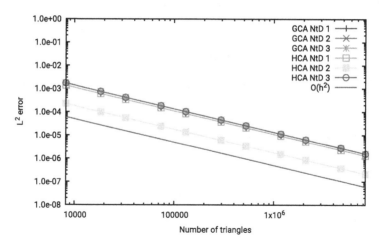

Fig. 3. L^2 error for the Neumann-to-Dirichlet problem

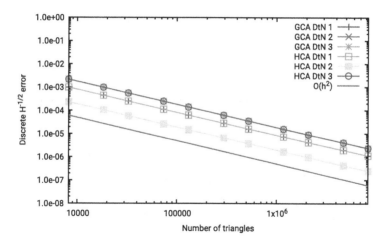

Fig. 4. Discrete $H^{-1/2}$ error for the Dirichlet-to-Neumann problem

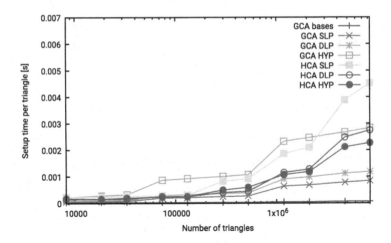

Fig. 5. Setup times for cluster bases and matrices

Since we expect the runtime to grow like $\mathcal{O}(n \log^\alpha n)$, cf. [9] in combination with [5, Lemma 3.2] for HCA and [7] in combination with [6, Lemma 3.45] for GCA, where n is the number of triangles and $\alpha > 0$, we display the runtime *per triangle* in Fig. 5, using a logarithmic scale for n and a linear scale for the runtime. We can see that HCA and GCA behave differently if the problem size grows: the runtime for HCA shows jumps when the order of interpolation is increased, while the runtime for GCA shows jumps when the order of the nearfield quadrature is increased.

This observation underlines a fundamental difference between the two methods: HCA constructs the full coefficient matrix S for every matrix block, and the matrix S requires m^6 coefficients to be stored and $\mathcal{O}(km^3)$ operations for the adaptive cross approximation, where we expect $k \sim m^2$. GCA, on the other hand, applies cross approximation only per cluster, not per block, and the recursive structure of the algorithm ensures that only indices used in a cluster's children are considered in their parent. This explains why GCA is less vulnerable to increases in the order than HCA.

On the other hand, HCA computes the approximation of a submatrix by evaluating *single* integrals, cf. (11), that can be computed in $\mathcal{O}(q_{\text{near}}^2)$ operations, while GCA relies on *double* integrals, i.e., the entries of the matrix G, that require $\mathcal{O}(q_{\text{near}}^4)$ operations. This explains why HCA is less vulnerable to increases in the nearfield quadrature order than GCA.

The setup of the matrices dominates the runtime, e.g., for 8 388 688 triangles, the matrices G, K, and W take 7 096, 9 937, and 23 752 s to set up with GCA, respectively, while the preconditioner for G takes only 4 898 s for coarsening and 2 244 s for the factorization, with 2 362 and 1 470 s for the preconditioner for W. Solving the linear system with the preconditioned conjugate gradient method takes around 750 s for the Dirichlet-to-Neumann problems and around 500 s for the Neumann-to-Dirichlet problems.

Of course, the storage requirements of the algorithms may be even more important than the runtime, since they determine the size of a problem that "fits" into a given computer. We again expect a growth like $\mathcal{O}(n \log^\alpha n)$ and report the storage requirements *per triangle* in Fig. 6.

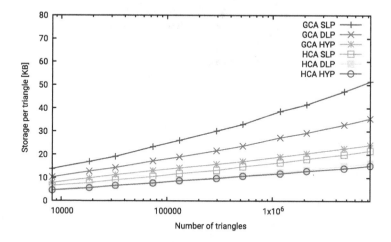

Fig. 6. Storage requirements for the matrices

Although theory would lead us to expect the storage requirements to grow like $\mathcal{O}(n \log^2 n)$, Fig. 6 suggests a behaviour more like $\mathcal{O}(n \log n)$ in practice. We can see that HCA consistently requires less storage than GCA. This is not surprising, since the algebraic algorithm [5] employed to turn the hierarchical matrix provided by HCA into an $\mathcal{H}^2$-matrix essentially computes the *best* possible $\mathcal{H}^2$-matrix approximation.

We may conclude that a server with 2 processors and just $48 = 2 \times 24$ processor cores equipped with 1 536 GB of main memory can handle boundary element problems with more than 8 million triangles in a matter of hours without sacrificing accuracy.

Admittedly, the unit sphere considered so far is an academic example. In order to demonstrate that the techniques also work in more complicated settings, we consider the crank shaft geometry displayed in Fig. 7 created by Joachim Schöberl's **netgen** Software. We start with a mesh with 25 744 triangles and refine these triangles regularly in order to obtain higher resolutions.

The boundary element mesh is far less "smooth" in this case, and this leads both to an increased condition number and the need to use significantly higher nearfield quadrature orders. Since we have already seen that HCA is far less susceptible to the nearfield quadrature than GCA, we only consider HCA in this example. Experiments indicate that the parameters given in Table 2 are sufficient to preserve the theoretically predicted convergence rates of the Galerkin method.

Fig. 7. Crank shaft geometry

Table 2. Parameters chosen for the crank shaft geometry

n	q_{near}	r_{leaf}	m	ϵ_{aca}	ϵ_{comp}	ϵ_{slv}	ϵ_{prc}
25 744	7	64	5	1_{-9}	1_{-9}	1_{-11}	1_{-3}
102 976	8	64	6	1_{-10}	1_{-10}	1_{-12}	5_{-4}
231 696	9	64	7	1_{-11}	1_{-11}	1_{-13}	2_{-4}
411 904	9	64	8	1_{-12}	1_{-12}	1_{-14}	1_{-4}
926 784	10	64	9	1_{-12}	1_{-12}	1_{-14}	5_{-5}
1 647 616	10	64	10	1_{-13}	1_{-13}	1_{-15}	5_{-5}

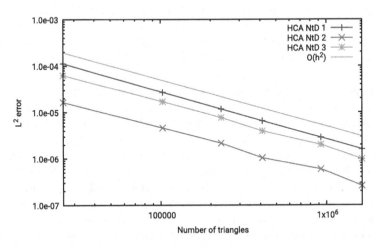

Fig. 8. L^2 error for the Neumann-to-Dirichlet problem for the crank shaft geometry

Figure 8 shows the L^2-norm errors for the Neumann-to-Dirichlet problem at different refinement levels of the crank shaft geometry. We can see that the optimal $\mathcal{O}(h^2)$ rate of convergence is again preserved despite the matrix compression.

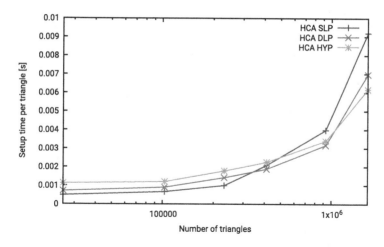

Fig. 9. Setup times for matrices for the crank shaft geometry

Figure 9 shows the setup times per triangle for the three matrices. Due to the computationally expensive nearfield quadrature, the setup time for the matrices dominates the other parts of the program, e.g., for 1 647 616 triangles the setup times for the three matrices are 15 167, 11 488, and 10 164 s, respectively, while computing both preconditioners takes only 3 156 s and each linear system is solved in under 180 s.

We conclude that using modern compression techniques like HCA and GCA in combination with efficient $\mathcal{H}^2$-matrix representations of the resulting matrices, large boundary element problems on meshes with several million triangles can be treated in few hours on moderately expensive servers.

References

1. Bebendorf, M.: Approximation of boundary element matrices. Numer. Math. **86**(4), 565–589 (2000)
2. Bebendorf, M.: Hierarchical LU decomposition based preconditioners for BEM. Computing **74**, 225–247 (2005)
3. Bebendorf, M., Kuske, C., Venn, R.: Wideband nested cross approximation for Helmholtz problems. Numer. Math. **130**(1), 1–34 (2014). https://doi.org/10.1007/s00211-014-0656-7
4. Bebendorf, M., Rjasanow, S.: Adaptive low-rank approximation of collocation matrices. Computing **70**(1), 1–24 (2003)
5. Börm, S.: Construction of data-sparse $\mathcal{H}^2$-matrices by hierarchical compression. SIAM J. Sci. Comp. **31**(3), 1820–1839 (2009)
6. Börm, S.: Efficient Numerical Methods for Non-local Operators: $\mathcal{H}^2$-Matrix Compression, Algorithms and Analysis, EMS Tracts in Mathematics, vol. 14, EMS (2010)
7. Börm, S., Christophersen, S.: Approximation of integral operators by Green quadrature and nested cross approximation. Numer. Math. **133**(3), 409–442 (2015). https://doi.org/10.1007/s00211-015-0757-y

8. Börm, S., Grasedyck, L.: Low-rank approximation of integral operators by inter-polation. Computing **72**, 325–332 (2004)

9. Börm, S., Grasedyck, L.: Hybrid cross approximation of integral operators. Numer. Math. **101**, 221–249 (2005)

10. Cohen, A., Dahmen, W., DeVore, R.: Adaptive wavelet methods for elliptic oper-ator equations – convergence rates. Math. Comp. **70**, 27–75 (2001)

11. Dahmen, W., Schneider, R.: Wavelets on manifolds I: construction and domain decomposition. SIAM J. Math. Anal. **31**, 184–230 (1999)

12. Erichsen, S., Sauter, S.A.: Efficient automatic quadrature in 3-d Galerkin BEM. Comput. Meth. Appl. Mech. Eng. **157**, 215–224 (1998)

13. Giebermann, K.: Multilevel approximation of boundary integral operators. Com-puting **67**, 183–207 (2001)

14. Grasedyck, L.: Adaptive recompression of $\mathcal{H}$-matrices for BEM. Computing **74**(3), 205–223 (2004)

15. Grasedyck, L., Hackbusch, W.: Construction and arithmetics of $\mathcal{H}$-matrices. Com-puting **70**, 295–334 (2003)

16. Greengard, L., Rokhlin, V.: A fast algorithm for particle simulations. J. Comp. Phys. **73**, 325–348 (1987)

17. Greengard, L., Rokhlin, V.: A new version of the fast multipole method for the Laplace equation in three dimensions. In: Acta Numerica 1997, pp. 229–269. Cam-bridge University Press (1997)

18. Hackbusch, W.: Regularity. Elliptic Differential Equations. SSCM, vol. 18, pp. 263–310. Springer, Heidelberg (2017). https://doi.org/10.1007/978-3-662-54961-2_9

19. Hackbusch, W.: A sparse matrix arithmetic based on $\mathcal{H}$-matrices. Part I: introduc-tion to $\mathcal{H}$-matrices. Computing **62**(2), 89–108 (1999)

20. Hackbusch, W.: Tensor Spaces. Hierarchical Matrices: Algorithms and Analysis. SSCM, vol. 49, pp. 379–394. Springer, Heidelberg (2015). https://doi.org/10.1007/978-3-662-47324-5_16

21. Hackbusch, W., Khoromskij, B.N., Sauter, S.A.: On $\mathcal{H}^2$-matrices. In: Bungartz, H., Hoppe, R., Zenger, C. (eds.) Lectures on Applied Mathematics, pp. 9–29. Springer-Verlag, Berlin (2000). https://doi.org/10.1007/978-3-642-59709-1_2

22. Harbrecht, H., Schneider, R.: Wavelet Galerkin schemes for boundary integral equations - implementation and quadrature. SIAM J. Sci. Comput. **27**, 1347–1370 (2006)

23. Langer, U., Pusch, D., Reitzinger, S.: Efficient preconditioners for boundary ele-ment matrices based on grey-box algebraic multigrid methods. Int. J. Numer. Meth. Eng. **58**(13), 1937–1953 (2003)

24. von Petersdorff, T., Schwab, C.: Fully discretized multiscale Galerkin BEM. In: Dahmen, W., Kurdila, A., Oswald, P. (eds.) Multiscale Wavelet Methods for PDEs, pp. 287–346. Academic Press, San Diego (1997)

25. Rokhlin, V.: Rapid solution of integral equations of classical potential theory. J. Comp. Phys. **60**, 187–207 (1985)

26. Sauter, S.A.: Cubature techniques for 3-D Galerkin BEM. In: Hackbusch, W., Wittum, G. (eds.) Boundary Elements: Implementation and Analysis of Advanced Algorithms. Notes on Numerical Fluid Mechanics (NNFM), vol. 50, pp. 29–44. Vieweg-Verlag, Berlin (1996). https://doi.org/10.1007/978-3-322-89941-5_2
27. Sauter, S.A., Schwab, C.: Boundary Element Methods. Springer, Cham (2011)
28. Steinbach, O., Wendland, W.L.: The construction of some efficient preconditioners in the boundary element method. Adv. Comp. Math. **9**, 191–216 (1998)
29. Tausch, J.: A variable order wavelet method for the sparse representation of layer potentials in the non-standard form. J. Numer. Math. **12**(3), 233–254 (2004)

Solving Large-Scale Interior Eigenvalue Problems to Investigate the Vibrational Properties of the Boson Peak Regime in Amorphous Materials

Giuseppe Accaputo[1], Peter M. Derlet[2], and Peter Arbenz[1(✉)] (iD)

[1] Computer Science Department, ETH Zürich, Zürich, Switzerland
arbenz@inf.ethz.ch
[2] Condensed Matter Theory Group, Paul Scherrer Institut, Villigen, Switzerland

Abstract. Amorphous solids, like metallic glasses, exhibit an excess of low frequency vibrational states reflecting the break-up of sound due to the strong structural disorder inherent to these materials. Referred to as the boson peak regime of frequencies, how the corresponding eigenmodes relate to the underlying atomic-scale disorder remains an active research topic. In this paper we investigate the use of a polynomial filtered eigensolver for the computation and study of low frequency eigenmodes of a Hessian matrix located in a specific interval close to the boson peak regime. A distributed-memory parallel implementation of a polynomial filtered eigensolver is presented. Our implementation, based on the Trilinos framework, is then applied to a Hessian matrix of an atomistic bulk metallic glass structure derived from a molecular dynamics simulation for the computation of eigenmodes close to the boson peak. In addition, we study the parallel scalability of our implementation on multicore nodes. Our resulting calculations successfully concur with previous atomistic results, and additionally demonstrate a broad cross-over of boson peak frequencies within which sound is seen to break-up.

Keywords: Amorphous materials · Boson peak · Large scale eigenvalue problems · Interior eigenvalues · Polynomial filters · Trilinos

1 Introduction

In amorphous materials, such as structural glass, sound waves have a meaning only within a finite range of wavelengths. At long wavelengths, the heterogeneity of the amorphous structure self averages [23] and an elastic continuum emerges. In this regime, sound is well defined via a linear dispersion characterized by a group velocity set by the continuum's isotropic elastic constants. However, as the wavelength reduces, the structural heterogeneity of the glass is increasingly probed, resulting in a broadening of a sound mode's frequency spectrum. When this broadening becomes comparable to the frequency of the sound wave (the Ioffe–Regel

© Springer Nature Switzerland AG 2021
T. Kozubek et al. (Eds.): HPCSE 2019, LNCS 12456, pp. 80–98, 2021.
https://doi.org/10.1007/978-3-030-67077-1_5

limit), sound loses its traditional meaning as a propagating plane wave. In this frequency range, the density of vibrational modes allowed by the solid is enhanced (the boson peak), also suggesting a transition from propagating to more localized or quasi-localized non-propagating modes. The precise way in which this transition is related to micro-structural length scales within the amorphous system remains an active area of research [11,15,25,27,28,35–37,39,49].

One avenue in which this phenomenon may be studied is via the molecular dynamics simulation technique – a particle trajectory method able to produce structural glasses with atomic scale resolution. Indeed computer generated amorphous structures may be generated in which the force on each atom is identically zero. When this is the case, the structure is at a local minimum of the total potential energy and its vibrational properties may be investigated through the corresponding local quadratic curvature. This is done by realizing that the leading order deviation in energy of a stable configuration (defined by the N atomic positions, r_i) may be expressed as the quadratic form

$$E\left(\{r_i + q_i\}\right) = E\left(\{r_i\}\right) + \frac{1}{2}\sum_{i,j=1}^{N} q_{ij}\Delta_{ij}q_{ij}, \tag{1}$$

where $q_{ij} = q_i - q_j$, and q_i is the deviation of the ith atom from its position r_i. In the above, Δ_{ij} is therefore the second derivative with respect to the bond-length deviations q_{ij}. Δ_{ij} is a symmetric 3×3 matrix. The quadratic energy term then defines a linear restoring force for such deviations, and an equation of motion for the $\{q_i\}$ coordinates whose secular form equals

$$\sum_{j=1}^{N} \left(\delta_{ij}(\omega_n)^2 - H_{ij}\right) q_{n,j} = 0, \qquad 1 \le n \le N,$$

where $H_{ij} = \Delta_{ij}/\sqrt{m_i m_j}$, m_i is the mass of the ith atom, and ω_n is the frequency of the nth vibrational eigenmode $q_{n,i}$. The energy function, $E\left(\{r_i\}\right)$, is usually determined through an empirical force model which is short range (for metallic and covalent systems), i.e., which spans a few atomic distances and results in a sparse matrix H.

Therefore, in order to calculate the vibrational frequencies, $\omega_n = \sqrt{\lambda_n}$, we have to solve a real symmetric eigenvalue problem

$$Hq = \lambda q, \qquad H \in \mathbb{R}^{3N \times 3N}, \qquad q \in \mathbb{R}^{3N}. \tag{2}$$

The regime of frequencies relevant to sound waves and the boson peak regime are at the lower end of the eigenvalue spectrum. Early simulation work had often considered system sizes in the range of several thousand atoms, and more recent work has considered system sizes up to several hundred thousand atoms [15,27,39]. Contemporary understanding of the frequency regime of the boson peak suggests that the relevant length scales correspond to those of elastic heterogeneities – a length scale which is at least an order of magnitude larger than an inter-atomic distance. Thus, if one wishes to study the transition from well defined propagating sound waves to their break up, larger system sizes will be needed spanning values of N up to several tens if not hundreds of millions of atoms. For such large system sizes, the boson peak eigenvalue regime is no longer an extremal eigenvalue problem, since there will now be a significant interval of (lower) eigenvalues describing the allowed sound waves. This fact motivates the development of eigensolver methods which are able to focus on a finite interval of eigenvalues in the *interior* of the spectrum of H and their eigenvectors.

The shift-and-invert Lanczos (SI-Lanczos) algorithm is the method of choice for computing interior eigenvalues and corresponding eigenvectors of a symmetric H close to some target τ. However, the SI-Lanczos algorithm needs the factorization of $H - \tau I$ which is not feasible here for its excessive memory requirements. For such cases, the Jacobi–Davidson methods have been developed [42,43]. To be efficient, they however need an effective preconditioner to solve the so-called correction equation, which usually entails its factorization [4]. In an earlier study [34], we were not able to identify such preconditioners for (2).

In this work we investigate a technique, known as *spectral filtering* or *kernel polynomial method*, for solving eigenvalue problems that obviates factorizations altogether [20,21,40,41]. Spectral filtering has been combined with Krylov space methods [16,38] or subspace iteration [18,51]. High-performance implementations are discussed in [18,31]. In order for the technique to be applicable the extremal eigenvalues $\lambda_{\min}$ and $\lambda_{\max}$ of H, or, at least, some accurate bounds must be available. To compute the eigenvalues in the interval $[\xi, \eta] \subset [\lambda_{\min}, \lambda_{\max}]$, a polynomial ρ is constructed such that $\rho(x) \geq 1$ in $[\xi, \eta]$ and $|\rho(x)| \ll 1$ away from $[\xi-\varepsilon, \eta+\varepsilon]$ for some positive ε. The desired polynomial ρ could be an approximation of the characteristic function $\chi_{[\xi,\eta]}$ associated with the interval $[\xi, \eta]$. If $\rho(H)$ multiplies a vector, (most) of the unwanted eigenvector components are suppressed. Therefore, ρ is called a *polynomial filter*. The degree of ρ depends on the width of the interval $[\xi, \eta]$, on the width ε of the margins, and the strength of the filter. The degree increases if $\eta - \xi$ and/or ε shrink. A consequence is that increasing parallelism by slicing the interval $[\xi, \eta]$ is not scalable. Interval slicing may however be necessary for memory reasons. In our experiments we use polynomial degrees as high as $d = \mathcal{O}(10'000)$.

The numerical experiments consider that part of the spectrum in which sound is known to break up in a simplified model of an amorphous solid corresponding to a Hessian H of order $4'116'000$ corresponding to $1'372'000$ atoms. With the new approach we can deal with models that are more than five times bigger than those we reported on previously. Their simulation requires at least 360 cores on our compute environment in order to store matrices and vectors in main memory. The new algorithm relies heavily on matrix-vector multiplication and therefore scales well to higher core counts.

For the amorphous system investigated in the present work we find a spectrum of low frequency modes that are well characterized by sound waves. However as the frequency of these modes increases, the associated decrease in sound wavelength results in increased scattering with the underlying microscopic disorder of the amorphous material until eventually the vibrational modes have little or no sound-like character. This is the boson peak regime and for the largest system size considered, the transition appears to have a rather extended frequency range suggesting the boson peak frequency and the associated break-up of sound is a broad cross-over rather than an abrupt transition.

The paper is organized as follows. In Sect. 2 we review the technique of polynomial filtering. In Sect. 3 we give some details on how we implemented our eigensolvers with the Trilinos software framework. In Sect. 4 we discuss the numerical experiments that we conducted in a distributed memory computing environment. In Sect. 5 we discuss physics results and in Sect. 6 we draw our conclusion and mention potential future work.

An extended version of this paper is available from the arXiv [2].

2   Numerical Solution Procedures

In this section we discuss how we complement the restarted Lanczos algorithm by polynomial filters to compute interior eigenvalues of a symmetric matrix. The Lanczos algorithm has been used for this purpose, e.g., in [9,24].

2.1   Spectral Projector

Let H be a real symmetric matrix of order n and let

$$H = U\Lambda U^* = \sum_{i=1}^{n} \lambda_i u_i u_i^*, \qquad U = [u_1, \ldots, u_n], \quad \Lambda = \mathrm{diag}(\lambda_1, \ldots, \lambda_n),$$

with orthogonal/unitary U, be its spectral decomposition. For convenience, we assume that H's spectrum $\sigma(H) \subset [-1, 1]$. If this is not the case then we can enforce this property by means of a linear transformation [2] provided that we know the extremal eigenvalues $\lambda_{\min}$ and $\lambda_{\max}$ of H or that we at least have accurate lower and upper bounds, respectively, for them.

To compute the eigenpairs associated with the eigenvalues in a prescribed interval $[\xi, \eta] \subset [-1, 1]$ it is useful to define the corresponding spectral projector. To this end, let

$$\chi_{[\xi,\eta]}(x) = \begin{cases} 1, & x \in [\xi, \eta], \\ 0, & \text{otherwise}, \end{cases} \tag{3}$$

be the *characteristic function* for the closed interval $[\xi, \eta]$. Then, the *spectral projector* [33] for the eigenvalues in $[\xi, \eta]$ is given by

$$P_{[\xi,\eta]} \equiv \chi_{[\xi,\eta]}(H) = \sum_{i=1}^{n} \chi_{[\xi,\eta]}(\lambda_i) u_i u_i^* = \sum_{\xi \le \lambda_i \le \eta} u_i u_i^*. \tag{4}$$

The orthogonal projector $P_{[\xi,\eta]}$ has eigenvalues 0 and 1. Its range $\mathcal{R}(P_{[\xi,\eta]})$ is spanned by the eigenvectors u_i with $\lambda_i \in [\xi, \eta]$. The trace of the projector $P_{[\xi,\eta]}$ is the number of eigenvalues in $[\xi, \eta]$, counting multiplicities,

$$\mu_{[\xi,\eta]} \equiv \mathrm{trace}\, P_{[\xi,\eta]} = |\sigma(H) \cap [\xi, \eta]|. \tag{5}$$

Algorithm 1. Computation of the eigenvectors associated with an interval

---

Input: Symmetric positive definite Matrix H with $-1 \lesssim \lambda_{\min}$ and $\lambda_{\max} \lesssim 1$ and an interval $[\xi, \eta] \subset [-1, 1]$.
Output: Eigenpairs $(\lambda_1, u_1), \ldots, (\lambda_m, u_m)$, $m = \mu_{[\xi,\eta]}$, with $\{\lambda_1, \ldots, \lambda_m\} = \sigma(H) \cap [\xi, \eta]$.

1: Determine an orthonormal basis $V = [v_1, \ldots, v_m]$ for $\mathcal{R}(P_{[\xi,\eta]})$.
2: Determine the desired eigenpairs by the Rayleigh–Ritz procedure [30], i.e., compute the spectral decomposition of the (small) matrix V^*HV,
$$Q^*(V^*HV)Q = \Lambda. \tag{6}$$
The eigenvalues in $[\xi, \eta]$ can now be read from the diagonal of Λ; the associated eigenvectors are the respective columns of $U = VQ$.

In Algorithm 1 we formulate an *idealized procedure* to compute the eigenvalues of H in $[\xi, \eta]$ and their corresponding eigenvectors. In step 1 of this algorithm it is useful to know (at least an upper bound of) the dimension $\mu_{[\xi,\eta]}$ of $\mathcal{R}(P_{[\xi,\eta]})$. Then, V can be computed by the Lanczos algorithm with the matrix $P_{[\xi,\eta]}$ in (4). Applying $P_{[\xi,\eta]}$ to a vector removes all components in the direction of the unwanted eigenvectors. Algorithm 1 implements an idealized procedure as the spectral projector $P_{[\xi,\eta]}$ is not available. Forming it would require the knowledge of the seeked eigenvectors. If $P_{[\xi,\eta]}$ was available the desired subspace could be obtained by one step of subspace iteration provided the subspace is chosen properly.

In a practical procedure, a function, say ψ, is constructed that is much bigger in $[\xi, \eta]$ than in $[-1, 1] \setminus [\xi, \eta]$ such that the components in undesired directions are suppressed as much as possible if $\psi(H)$ is applied to a vector. It is not necessary that $\psi(H) \approx \chi_{[\xi,\eta]}(H)$.

In the sequel we discuss *polynomial filters* ψ. We favor polynomial filters since applying a matrix polynomial to a vector requires matrix-vector multiplications which can be implemented relatively easily and efficiently in an HPC environment. Rational approximations are possible but require the solution of linear systems which we want to avoid [7,50]. In fact, the efficient implementation of the multiplication of sparse matrices with (multiple) vectors is an area of recent investigation [3,22].

2.2   Chebyshev Polynomial Expansions

Let $\mathbb{P}_j$ denote the set of polynomials of degree at most j. The Chebyshev polynomials $T_j(x) = \cos(j \arccos x) \equiv \cos(j\vartheta) \in \mathbb{P}_j$, $j = 0, 1, \ldots$, form a complete orthogonal set on the interval $[-1, 1]$ with respect to the inner product [44]

$$\langle f, g \rangle \equiv \int_{-1}^{1} \frac{f(x)g(x)}{\sqrt{1-x^2}} \, dx = \int_{0}^{\pi} f(\vartheta)g(\vartheta) \, d\vartheta, \qquad x = \cos \vartheta.$$

Using the Kronecker delta δ_{jk}, we have

$$\langle T_j, T_k \rangle = \frac{\pi}{2}(1 + \delta_{0j})\delta_{jk}.$$

The Chebyshev series of a piecewise continuous function f on $[-1, 1]$ is given by

$$\hat{f}(x) = \sum_{j=0}^{\infty} \gamma_j T_j(x), \qquad \gamma_j = \frac{\langle f, T_j \rangle}{\langle T_j, T_j \rangle}. \tag{7}$$

This series converges to $f(x)$ if f is continuous at the point x, and converges to the average of the left- and right-hand limits if f has a jump discontinuity at x [44]. The polynomial $p \in \mathbb{P}_d$ that best approximates f in the norm $\|\cdot\| = \langle \cdot, \cdot \rangle^{1/2}$ is obtained by *truncation*,

$$p_d(x) = \sum_{j=0}^{d} \gamma_j T_j(x). \tag{8}$$

The Chebyshev polynomials satisfy the three-term recurrence

$$T_{j+1}(x) = 2x T_j(x) - T_{j-1}(x), \ j > 0, \quad T_0(x) = 1, \ T_1(x) = x.$$

The three-term recurrence can be used when a matrix polynomial is applied to a vector. Let $t_j = T_j(H)x$. Then $t_0 = T_0(H)x = Ix$, $t_1 = T_1(H)x = Hx$, and

$$t_{j+1} = 2H t_j - t_{j-1}, \qquad j > 0. \tag{9}$$

Algorithm 2 shows how $p_d(H)x$ is evaluated employing the 3-term recurrence (9). It constitutes the most time consuming operation in our simulations. Note that the degree d can be in the hundreds or even thousands. Algorithm 2 presents a stable procedure to evaluate truncated Chebyshev series [44]. (The Clenshaw algorithm [44] could be used as well.)

2.3 Dealing with the Gibbs Phenomenon

Truncated Chebyshev series expansions of discontinuous functions exhibit oscillations near the discontinuities, which are known as *Gibbs oscillations* or *Gibbs phenomenon*. To alleviate this phenomenon the series must be truncated smoothly using appropriate damping factors. The damping factors depend on the

---

Algorithm 2. Evaluation of truncated Chebyshev series $p_d(H)x$

Input: Vector x and coefficients $\gamma_0, \dots, \gamma_d$ that define p_d in (8).
Output: Vector $y = p_d(H)x$.

1: $t'' = x$; $y = \gamma_0 t''$.   /* $t'' = t_0$; $y = p_0(H)x$. */
2: if $d \geq 1$ then $t' = Hx$; $y = y + \gamma_1 t'$; end if   /* $t' = t_1$; $y = p_1(H)x$. */
3: for $j = 2, \dots, d$ do
4:    $t = 2H t' - t''$; $t'' = t'$; $t' = t$.   /* $t = t_j$; $t' = t_{j-1}$; $t'' = t_{j-2}$. */
5:    $y = y + \gamma_j t$.   /* $y = p_j(H)x = p_{j-1}(H)x + \gamma_j t_j$. */
6: end for

index at which the series is truncated, i.e., on the degree of the approximating polynomial. Thus, $p_d(t)$ in (8) is replaced by

$$\rho_d(t) = \sum_{j=0}^{d} g_j^d \gamma_j T_j(t) \tag{10}$$

where the g_j^d are the smoothing coefficients. These coefficients can be determined by different approaches, see [47] for a survey. We chose to employ Jackson smoothing [32,38], with smoothing coefficients given by

$$g_j^d = \left(1 - \frac{j}{d+2}\right)\cos j\alpha_d + \frac{1}{d+2}\frac{\cos\alpha_d}{\sin\alpha_d}\sin j\alpha_d, \qquad \alpha_d = \frac{\pi}{d+2}. \tag{11}$$

The advantage of Jackson smoothing is its monotonic approximation. This implies that the truncated polynomial is positive if the function to be approximated is so.

2.4   Counting the Eigenvalues in an Interval

In order that the eigenvalues of H in $[\xi, \eta]$ can be computed numerically their number or at least a (tight) upper bound has to be known. After all, memory space has to be provided for storing the associated eigenvectors. Applying Sylvester's law of inertia for counting the eigenvalues in an interval is infeasible because of the fill-in generated in the LDL^T factorization of $H - \lambda I$. However, we showed above that the number of the eigenvalues in an interval $[\xi, \eta]$ equals the trace of the spectral projector $P_{[\xi,\eta]} = \chi_{[\xi,\eta]}(H)$ which we do not have available explicitly, but which we can approximate by a truncated Chebyshev series $\psi(H)$, i.e. $\psi(t) \approx \chi_{[\xi,\eta]}(t)$. Hutchinson [19] showed that $\mathbb{E}(v^* H v) = \text{trace}(H)$ holds for randomly generated vectors v with entries that are identically independently distributed (i.i.d.) random variables. Hutchinson originally used i.i.d. Rademacher random variables, where each entry of v assumes the value -1 or 1 with probability $1/2$. In general, any sequence of random vectors v_ℓ whose entries are i.i.d. random variables can be used, as long as the mean of their entries is zero [9]. Here, we use a Gaussian estimator to approximate trace $P_{[\xi,\eta]}$,

$$\mu_{[\xi,\eta]} = \text{trace } P_{[\xi,\eta]} \approx T_M \equiv \frac{n}{M}\sum_{\ell=1}^{M} v_\ell^T \psi_d(H)v_\ell, \qquad \|v_\ell\|_2 = 1, \tag{12}$$

by using normally distributed variables for the entries of the random vectors v_ℓ. (The factor n in (12) is due to the normalization of the v_ℓ.) Despite the fact that the Gaussian estimator has a larger variance than Hutchinson's, it shows better convergence in terms of the number M of sample vectors [5]. As in [20,26,29] we choose $\psi_d \in \mathbb{P}_d$ to be a truncated Chebyshev series for $\chi_{[\xi,\eta]}(t)$, such that

$$\gamma_j = \frac{\langle \chi_{[\xi,\eta]}, T_j \rangle}{\langle T_j, T_j \rangle}.$$

2.5   Computing a Basis of $\mathcal{R}(P_{[\xi,\eta]})$

The desired eigenvectors u_k of H with eigenvalues $\lambda_k \in [\xi,\eta]$ span $\mathcal{R}(P_{[\xi,\eta]}) = \mathcal{R}(\chi_{[\xi,\eta]}(H))$. In Algorithm 1 first a basis for $\mathcal{R}(P_{[\xi,\eta]})$ is computed and then the eigenvectors are extracted from it by the Rayleigh–Ritz procedure [30]. Remember that if $u_k \in \mathcal{R}(V)$ then λ_k is an eigenvalue of V^*HV.

The procedure to generate a basis of $\mathcal{R}(P_{[\xi,\eta]})$ is based on the thick-restart Lanczos algorithm [48] where the operator is the matrix polynomial $\rho_d(H)$. Our implementation follows closely the one described by Li et al. [24] that has been implemented in the EVSL library[1].

The requirements for the polynomial filter are different for the Lanczos algorithm and for eigenvalue counting. In the latter the filter $\psi(t) \in \mathbb{P}_d$ has to be a good approximation of the characteristic function $\chi_{[\xi,\eta]}(t)$. As the Lanczos algorithm converges best towards extremal eigenvalues that are well separated from the rest of the spectrum [30], $\rho_d(t)$ must be (relatively) large in $[\xi,\eta]$ and small outside. Li et al. [24] suggest, as others before [40,41,47], to generate a filter that mimicks a Delta distribution, i.e., the functional $\delta(\cdot - \gamma)$ defined by

$$\int_\infty^\infty \delta(t - \gamma)\phi(t)\,dt = \phi(\gamma)$$

for all continuous functions ϕ. In $\mathbb{P}_d$, $\delta(t - \gamma)$ can be represented by

$$\rho_d(x) = \sum_{j=0}^d \frac{T_j(\gamma)}{\langle T_j, T_j \rangle} T_j(x). \tag{13}$$

$\gamma \in (\xi,\eta)$ is chosen close to the interval midpoint such that $\tau := \rho_d(\xi) = \rho_d(\eta)$. By construction, $\rho_d(x) > \tau$ in (ξ,η). Eigenvectors of $\rho_d(H)$ corresponding to eigenvalues $> \tau$ are potential eigenvectors of H. Care has to be taken, though, as different eigenvalues of H may be mapped to the same value by ρ_d. Nevertheless, the eigenvectors of $\rho_d(H)$ corresponding to eigenvalues $> \tau$ do span $\mathcal{R}(P_{[\xi,\eta]})$. The correct eigenvalue-eigenvector relations can be found by the Rayleigh–Ritz procedure applied to H. With this filter the eigenvalues close to ξ and η appear usually later than those close to γ. Since $\mathcal{T}_M$ in (12) is only an approximation of $\mu_{[\xi,\eta]}$ we add some 10% to it to get a (heuristic) upper bound for the number of eigenvalues in $[\xi,\eta]$. Of course, a large overestimation of $\mu_{[\xi,\eta]}$ entails a waste of memory space.

3   Implementation

We combined the methods discussed in the previous section and a few other useful tools into a utility that can be employed to compute the eigenpairs of a $n \times n$ real symmetric (or complex Hermitian) matrix H within a specified interval of interest $[\xi,\eta]$ in parallel by simply providing an XML configuration file [1]. The outline of the utility is displayed in Algorithm 3. The utility is written in C++11 and uses Trilinos [45] extensively. Trilinos[2] is a collection of open-source

---

[1] http://www-users.cs.umn.edu/~saad/software/EVSL/.
[2] https://trilinos.org/packages/.

Algorithm 3. The BosonPeak Utility

1: Import user-specified configuration via XML file.
2: Import matrix H.
3: **if** *requested* **then** estimate extremal eigenvalues $\lambda_{min}, \lambda_{max}$ of H using a few Lanczos steps.
4: Transform the matrix H such that $-1 \leq \lambda_{min} < \lambda_{max} \leq 1$.
5: **if** *requested* **then** estimate the number of eigenvalues in the specified interval $[\xi, \eta]$.
6: Compute the polynomial filter ρ_d.
7: Compute the eigenpairs (λ_k, u_k), $\|u_k\| = 1$, of H with $\lambda_k \in [\xi, \eta]$ and residual norms $r_k = \|H u_k - \lambda_k u_k\| < \epsilon$ using the thick restart Lanczos algorithm.
8: **if** *requested* **then** save eigenvalues λ_k, associated eigenvectors u_k, and residual norms r_k to disk.

---

software libraries, called *packages*, for the development of scientific applications. More information on the utility is found in [1,2].

The most basic Trilinos package is Epetra that provides classes for the construction and use of sequential and distributed parallel linear algebra objects. The Trilinos solver packages are designed to work with Epetra objects. The most used linear algebra objects in our implementation are (i) sparse matrices stored as `Epetra_CrsMatrix` objects in the compressed row storage (CRS) scheme, and (ii) `Epetra_MultiVector` objects that represent *multivectors*, i.e., collections of dense vectors. Each vector in an `Epetra_MultiVector` object is stored as a contiguous array of double-precision numbers. Both objects are extensively used for sparse matrix-vector multiplications in the various Trilinos solver packages. All Trilinos packages resort to a method called `Epetra_Operator::Apply` to multiply a matrix with a (multi)vector. Our operators are mostly matrix polynomials, and a call to `Epetra_Operator::Apply` entails the invocation of Algorithm 2.

Anasazi [6] is a package that offers a collection of algorithms for solving large-scale eigenvalue problems. As part of the package it provides solver managers to implement strategies for that purpose. We employ Anasazi's block Krylov–Schur eigensolver with thick restarts. The subspace iteration that we discuss below is not a part of Anasazi. We implemented it ourselves, based on `Epetra` data structures.

The *Teuchos* package is a collection of common tools used throughout Trilinos. Among other things, it provides templated access to BLAS and LAPACK interfaces, parameter lists that allow to specify parameters for different packages, and memory management tools for aiding in correct allocation and deletion of memory.

Part of Teuchos' memory management tools is an implementation of a smart Reference-Counted Pointer (RCP) class, which for an object tracks a count of the number of references to it held by other objects. Once the counter becomes zero, the object can be deleted. The advantage of a RCP is that the possibility of memory leaks in a program can be reduced. This is important when working with rather large objects, e.g., an `Epetra_CrsMatrix` object storing over 10^9 nonzero entries. RCP objects are heavily used throughout our implementation to manage large objects, especially large temporary objects that are only needed during a fraction of the whole computation.

Trilinos supports distributed-memory parallel computations through the Message Passing Interface (MPI). Both `Epetra_CrsMatrix` and `Epetra_Multivector` objects can be used in a distributed memory environment by defining data distribution patterns using `Epetra_Map` objects.

The entries of a distributed object (rows or columns of an `Epetra_CrsMatrix` or rows of an `Epetra_Multivector`) are represented by *global indices* uniquely over the entire object. A map essentially assigns global indices to available MPI ranks, which in our case corresponds to a single core of a processor.

For the addressing, local and global indices in Epetra use by default the 32-bit `int` type. Since our implementation is based on the C++11 language standard and we want to allow computations with large matrices, we explicitly use 64-bit global indices of type `long long` when working with distributed linear algebra objects. (Local indices are of type `int`.)

An `Epetra_Map` object encapsulates the details of distributing data over MPI ranks. In our implementation, we use *contiguous* and *one-to-one* maps for the block row-wise distribution of the `Epetra_MultiVector` objects. *Contiguous* means that the list of global indices on each MPI rank forms an interval and is strictly increasing. A *one-to-one* map allows a global index to be owned by a single rank. For the columns, the distribution pattern we are using distributes the complete set of global column indices for a given global row, meaning that if a rank p owns the global row index i, it also owns all global column indices j on that row, thus having local access to the global entry (i, j). The map used for the distribution of the columns is thus not a one-to-one map, since a global column index can be owned by multiple ranks.

The *matrix import* implemented in the utility allows to efficiently import large matrices stored in a HDF5[3] file directly to an `Epetra_CrsMatrix` object. HDF5 is a data model, library, and file format for storing and managing data collections of all sizes and complexity. One of the advantages of using the HDF5 file format to store and import large matrices is the possibility to use MPI to read the HDF5 files in parallel. For this reason Trilinos provides the `EpetraExt::HDF5` class for importing a matrix stored in a HDF5 file to a `Epetra_CrsMatrix`.

Since the `EpetraExt::HDF5` class currently does not provide an import function for matrices with 64-bit global indices of type `long long`, we extended the class by this functionality. The BosonPeak utility also provides a Python script that can be used to convert matrices stored in the MatrixMarket format[4] to a HDF5 file suitable for import. The utility is described in detail in [1].

---

[3] https://portal.hdfgroup.org/.

[4] https://math.nist.gov/MatrixMarket/.

4   Numerical Experiments

To test our code we consider a glassy structure comprising of $1'372'000$ atoms. The Hessian matrix H has order $n = 3 \times 1'372'000 = 4'116'000$ and $nnz = 1'028'329'164$ nonzero entries. The number of nonzeros per row is about 250, leading to a sparsity of $6 \cdot 10^{-5}$. The atomic positions of this structure are produced through a series of molecular dynamics simulations involving: a well-equilibrated liquid at temperatures well above the melting temperature, a quench to the lower temperatures of the amorphous solid regime, and a final relaxation which brings the system to a local potential energy minimum from which the dynamical matrix H can be calculated. Note that different equivalent initial distributions of the atoms lead to different *realizations* of the configuration. The empirical atomic interaction model used to perform these simulations is based on a Lennard–Jones force model [46], which describes the interaction between atoms of two types differing in both size and mass. Periodic boundary conditions are used to remove the explicit structural effect of a surface. For the chosen density, the periodicity length is $L = 101.714585$ where the unit distance is close to the average atomic bond length. Details of the sample preparation procedure and the resulting glassy structures may be found in Refs. [12–14].

Past work considering much smaller glassy atomic configurations using the Lennard–Jones potential [1,2,15] suggests that the eigenvalue regime at which sound breaks up – the so-called boson peak regime – occurs in the approximate λ-interval [1,2], where λ_{max} was around but above 1900. Since the infinite-dimensional operator underlying (2) is bounded, we do not expect λ_{max} to increase much with increased system size. Indeed, the actual matrix H has its eigenvalues in the interval $[0, 1941]$. (In the computation this interval is mapped to $[-1, 1]$, cf. [2].)

A survey of results for the partial eigenvalue computations of the Hessian H in five subintervals of $[\xi, \eta] = [1, 2]$ is given in Table 1. There are three intervals of length 0.014 with about 50 eigenvalues and two intervals of length 0.028 with about 105 eigenvalues. These computations were part of an exploration of the eigenstructure of H in the entire interval [1, 2]. For each interval $[\xi, \eta]$ we give the true, estimated, and requested numbers of eigenvalues in the respective interval. The true numbers of eigenvalues are obtained as a result of the computations that targeted at the requested numbers of eigenvalues in the intervals. These are $10 - 20\%$ more eigenvalues than estimated and, as such, a crude upper bound for $\mu_{[\xi,\eta]}$. The estimates have been obtained by the technique discussed in Subsect. 2.4 with $M = 30$ samples in (12) and degree $k = 100$ for which parameters de Napoli et al. [29] report very good results.

We computed the eigenvalues of H with Anasazi's block Krylov–Schur (BKS) algorithm (in fact, the thick restarted block Lanczos algorithm) with the same parameter values. The block size b was fixed at 8. The maximal dimension of the Krylov space $\mathcal{K}$ was set to $3 \times n_{ev}$, where n_{ev} is the number of requested eigenvalues. The number of blocks was thus limited to $\frac{3 \times n_{ev}}{\text{block size}}$. If the maximal dimension of the Krylov space is attained then a restart is issued. The threshold in the discussion after Eq. (13) was set to $\tau = 0.9$.

Table 1. Computational results for large glassy structure. ξ, η are given with reference to the interval $[0, 1941]$. The number of matrix-vector multiplications (#MatVec) equals (blk size) $\times$ (#blk steps) $\times$ (poly degree). Times are for 360 cores on Euler V.

blk size	#blks	max dim $\mathcal{K}$	re-starts	#blk steps	poly degree	#MatVec's	ξ	η	evs est	evs req	evs true	time [sec]
8	21	168	3	63	10'142	5'111'568	1.169	1.183	49	55	46	22'667
8	21	168	3	77	10'202	6'284'432	1.183	1.197	49	55	48	27'959
8	21	168	4	91	10'262	7'470'736	1.197	1.211	50	55	54	32'937
8	48	384	2	80	5'175	3'312'000	1.211	1.239	101	128	106	14'647
8	48	384	2	80	5'233	3'349'120	1.239	1.267	103	128	105	14'850

We have implemented the operator $\rho_d(\boldsymbol{H})$ in Trilinos. The degree d of the filter polynomial is given in Table 1. One call of the operator amounts to the execution of d matrix-multivector multiplications. Remember that $b = 8$. The bd MatVec's constitute more than 99% of the execution time of the solver.

The computations were carried out on Euler V of ETH Zurich's compute cluster[5]. Euler V contains 352 compute nodes (Hewlett-Packard BL460c Gen10), each equipped with two 12-core Intel Xeon Gold 5118 processors (2.3 GHz nominal, 3.2 GHz peak) and 96 GB of DDR4 memory clocked at 2.4 GHz. The nodes are connected via a 100 Gb/s InfiniBand EDR network. In the specified environment, our implementation worked with OpenMPI 1.65, HDF5 1.8.12, Boost 1.57.0, and Trilinos 12.2.1. Further, the code has been compiled with GCC 4.8.2 and the following optimization flags:

```
-ftree-vectorize -march=corei7-avx -mavx -std=c++11 -O3.
```

The convergence criterion requires that the residual norms

$$\|\rho_d(\boldsymbol{H})\boldsymbol{u}_k - \mu_k\boldsymbol{u}_k\| < \epsilon, \qquad \epsilon = 10^{-6}, \qquad \|\boldsymbol{u}_k\| = 1,$$

for all *requested* eigenpairs $(\mu_k, \boldsymbol{u}_k)$. This lead to very accurate eigenpairs of the original matrices,

$$r_k \equiv \|\boldsymbol{H}\boldsymbol{u}_k - \lambda_k\boldsymbol{u}_k\| < 3.5 \cdot 10^{-8}, \qquad \|\boldsymbol{u}_k\| = 1,$$

or better for all desired (true) eigenpairs $(\lambda_k, \boldsymbol{u}_k)$. Since we compute too many eigenpairs, the desired ones are finally too accurate. So, it is important for good efficiency to have accurate estimates for the eigenvalue counts. A high security margin entails high computational and memory costs, together with overly accurate results.

Interestingly, it is faster to compute the 105 eigenvalues in the longer intervals than the 50 eigenvalues in the shorter ones, essentially because of the lower degrees of the filter polynomials. Also the maximal dimension of the search space is relatively larger. This eases the extraction of the desired information. The overhead due to reorthogonalizations is negligible. The times in Table 1 are

---

[5] https://scicomp.ethz.ch/wiki/Euler.

Table 2. Execution times for $p = 360$ and $p = 720$ cores for the intervals in Table 1 and derived speedups.

evs conf	evs found	deg	$p = 360$		$p = 720$		'speedup'
			#blk steps	time	#blk steps	time	
55	46	10'142	63	22'667	63	11'570	1.96
55	48	10'202	77	27'959	63	11'506	2.43
55	54	10'262	91	32'937	77	14'229	2.31
128	106	5'175	80	14'647	112	10'570	1.39
128	105	5'233	80	14'850	112	10'547	1.41

very well in proportion with the number of MatVec's. We consistently observe 226 MatVec's per sec. The times given are the fastest out of three runs. This number of MatVec's per second amounts to ~1.29 Gflop/s per core which is about 3.5% of the nominal peak performance of 36.8 Gflop/s per core [17, p. 55]. This percentage is comparable with those listed in the High Performance Conjugate Gradients (HPCG) Benchmark project[6].

Arithmetic intensity refers to the number of floating point operations performed per amount of bytes moved. The arithmetic intensity of a MatVec of a sparse matrix stored in CRS format [8] with a multivector with b columns is

$$ai = \frac{2 \cdot b \cdot nnz}{12 \cdot nnz + (2 \cdot 8 \cdot b + 4) \cdot n}. \tag{14}$$

The coefficient 12 refers to the 8-byte values and the 4-byte column indices of the matrix. The $4n$ bytes are consumed by the row pointers. The arithmetic intensity for $b = 8$ is 1.28, while it is 0.17 for $b = 1$. Note that multivectors are read and written.

The STREAM benchmark[7] measures the sustainable memory bandwidth in high performance computers. Relevant for the MatVec is the triad operation $x \leftarrow y + \alpha z$. On an Intel Xeon Gold 5118, about 6.42 GB/s are moved per core in the STREAM TRIAD benchmark [17]. The performance of memory bound operations equals the arithmetic intensity times the memory bandwidth. In this way we get a performance of 1.06 Gflop/s for $b = 1$ and of 8.20 Gflop/s for $b = 8$. Our code thus performs rather as a single vector MatVec than as a multivector MatVec with eight columns.

In Table 2 we replicate the execution times of Table 1 for 360 cores and complement them with the execution times obtained with 720 cores. 360 cores correspond to 15 nodes; 720 cores correspond to 30 nodes of Euler V. Evidently, the runs with 720 cores should take only half the time of the ones on 360 cores, amounting to a speedup of two. Since most of the computing time is spent in (blocked) matrix-vector multiplications a speedup close to two can indeed be

---

[6] https://www.hpcg-benchmark.org/.
[7] https://www.cs.virginia.edu/stream/.

expected. The data is distributed among the cores in the standard block row-wise fashion of Trilinos. The computations are similar, in particular, the initial vectors are equal. Nevertheless, the number of iteration steps until convergence can differ significantly. Therefore, the speedup appears to be erratic. However, if we compare the execution times of the single block steps, then we observe speedups close to two, more precisely between 1.94 and 1.988. With 720 cores about 443 MatVec's are executed per second. In this sense, our code scales well.

5   Physics Results

In what follows, the eigenmode q is seen as consisting of N 3-vectors, $q_i = (q_i^1, q_i^2, q_i^3)$, where the superscript indicates the coordinate direction. To gain an estimate of the number of atoms involved in a normalized eigenmode the participation ratio [10] is used,

$$\text{PR} = \frac{1}{\sum_i |q_i|^4} = \frac{1}{\|q\|_4^4}, \tag{15}$$

where $\|q\|_2 = 1$ is assumed. When an eigenmode has constant values, say $|q_i| = 1/\sqrt{N}$ then PR $= 1$ and all atoms are said to partake in the eigenmode. On the other hand when $|q_i| = 1$ for the ith atom and $|q_j| = 0$ for all other atoms $j \neq i$, then PR $= 1/N$. For a plane (sound) wave of wave-vector k we have $q_i = \sqrt{2/N}\, \hat{\zeta}_q \sin k \cdot q_i$ entailing a PR $= 2/3$. Here $\hat{\zeta}_q$ is termed the polarization vector and is usually taken as being perpendicular (transverse sound) or parallel (longitudinal sound) to k.

Figure 1 plots the participation ratio of the entire spectrum considered in the present work. At $\lambda = 0$, there exist three modes with a participation ratio equal to unity, which correspond to the translational modes of the dynamical matrix. For the region up to approximately, $\lambda = 0.5$, the eigenvalues are seen to bunch into clusters with a participation ratio of approximately $2/3$, indicating plane-wave-like eigenmodes and the presence of well defined sound. The observed bunching and their multiplicity arise from a combination of the polarization vectors, $\hat{\zeta}_k$, and the allowed wave-vectors, $k_{[mnl]} = 2\pi/L(m, n, l)$. Here L is the periodicity length of the amorphous configuration, and m, n, and l are integers defining the allowed wave-lengths. Through inspection of the corresponding eigenmodes, a wave-vector family and polarization type can be identified for each bunching and are indicated in Fig. 1. As the participation drops with increasing eigenvalue magnitude, this identification process becomes more difficult with each peak (now significantly broadened) being well described by a range of different plane wave components.

Figure 2 displays the spatial structure of two such eigenmodes. In this figure, (a) demonstrates a mode that is well described by [100] transverse plane waves, and (b) a mode well described by [311] transverse plane waves. In both (a) and (b), three spatial structures are shown, where the left-most figure plots the atoms at their spatial coordinates colored according to $|q_i|^4$ derived from the actual eigenmode and the central figure plots their color according to $|q_i^{\text{PW}}|^4$

derived from the plane wave (PW) representation. The right-most figure plots only those atoms for which $|q_i|^4 > \max\{|q_i^{PW}|^4; i = 1, n\}$. Inspection of the left and central panels of Fig. 2 demonstrates that a large part of the eigenmode derived from the dynamical matrix is well described by a PW decomposition. On the other hand, the right most panels clearly show that their exist local regions of oscillator strength which are not described by the PW picture. For the longer wavelength [110] mode, Fig. 2a, these regions are rare, but as the wavelength decreases (wave-vector magnitude increases), such as the [311] mode in Fig. 2b these localized regions become more numerous and somewhat extended. For higher wave-vector magnitudes, this trend continues with an associated drop in the participation ratio corresponding to the final break up of sound.

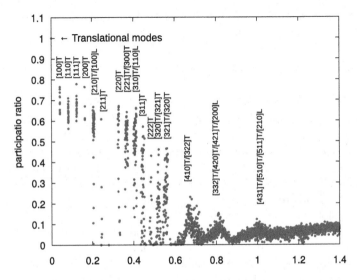

Fig. 1. The participation ratio (15) as a function of eigenvalue λ. At $\lambda = 0$, three uniform translational modes exist, having a PR equal to unity. For increasing values of λ, a bunching of eigenvalues is observed, all of which initially have a PR $= 2/3$, and correspond to eigenmodes which are well described by the plane wave representation $[m, n, l] \Leftrightarrow k_{mnl} = 2\pi/L(m, n, l)$ of either transverse (T) or longitudinal (L) polarization.

Via Fig. 1, both an eigenvalue and wave-vector magnitude regime can be identified at which the participation ratio rapidly decreases. The large system size presently considered allows us to study this regime in detail, suggesting that a crossover to more heterogeneous modes occurs over a broad range of eigenvalues. This corresponds to length scales of the order of $2\pi/|k_{410}|$ to $2\pi/|k_{332}|$ and length scales ranging between 20 and 25 bond lengths. Such a length-scale is compatible with amorphous elastic heterogeneity – a length scale which is believed to play a defining role in the break up of sound [27,35]. Larger system sizes will be needed to investigate whether this cross-over limits to a sharp transition at a distinct length-scale and particular eigenvalue that may be finally identified as the boson peak frequency. It is in such future simulations, that the true power of the current method will become evident since all computational resources can now be focused to the actual eigenvalue region of interest centered around the boson peak frequency.

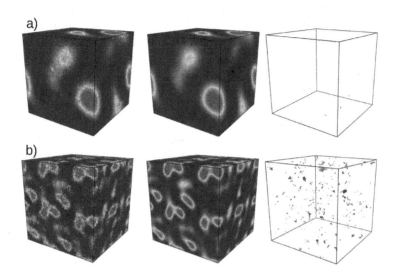

Fig. 2. Plots of the amorphous configuration consisting of 1372000 atoms, where each atom is colored according to its value of $|q_i|^4$. The left most panels derive its value from the calculated eigenmode and central panels from the plane wave decomposition. The right-most panels show only those atoms whose oscillator strength significantly deviates from the plane wave decomposition. (a) shows a [110] transverse mode, and (b) a [311] transverse mode. (Color figure online)

6   Conclusions

We have discussed a highly parallel implementation of a polynomial filtered Krylov space-based method for solving large-scale symmetric eigenvalue problems arising in the investigation of amorphous materials. The algorithm enables us to compute hundreds or thousands of eigenvalues of matrices of size in the millions. If the number of eigenvalues is too large to compute at once (e.g. for memory requirement) then the interval of interest can be split in subintervals that contain a reasonable number of eigenvalues whose associated eigenvectors can be accommodated by the available memory.

The polynomial filter is designed to enhance the eigenvalues of the interval of interest. Polynomials of very high degrees can result. Therefore, our algorithm is based almost completely on matrix-vector multiplications. The work to keep the basis vectors of the Krylov space orthogonal is negligible. This entails a high potential for parallelization which is confirmed by our experiments.

We plan to apply our solver to problems of size $49'152'000$ and larger. The use of larger matrices and correspondingly larger simulation cell sizes will give information on how the vibrational modes of the boson peak regime observed in the present work evolve to the bulk limit. Indeed, the treatment of larger system sizes will result in a transition to a spectrum free of eigenvalue bunching, where the effect of disorder smears the allowed sound-waves into an effectively continuous eigenvalue spectrum. In this experimentally relevant limit, the bulk nature of the boson peak regime should become manifest from an entirely atomistic description of a model amorphous system.

We will also work on optimizing the evaluation of matrix polynomials, in particular, if applied to multivectors, cf. [3,22]. As noted in the introduction, A is a matrix formed of *symmetric* 3×3 blocks. In an efficient implementation this should be taken into account. Doing so, the arithmetic intensity is more than doubled relative to the standard elementwise CRS storage format [8] and execution times can be reduced by about the same factor. (The numerator in (14) becomes $(8 \cdot \frac{2}{3} + 4 \cdot \frac{1}{9})nnz + (16 \cdot b + 4)n$.) Finally, let us note that the abundance of matrix-vector multiplications makes our code amenable to GPU computing.

Acknowledgments. The computations have been executed on the Euler compute cluster at ETH Zurich at the expense of a grant of the Seminar for Applied Mathematics. We acknowledge the assistance of the Euler Cluster Support Team.

References

1. Accaputo, G.: Solving large scale eigenvalue problems in amorphous materials. Master's thesis, ETH Zurich, Computer Science Department (2017). https://doi.org/10.3929/ethz-b-000221499
2. Accaputo, G., Derlet, P.M., Arbenz, P.: Solving large-scale interior eigenvalue problems to investigate the vibrational properties of the boson peak regime in amorphous materials. Print Archive: arXiv:1902.07041 [physics.comp-ph] (2019)
3. Aktulga, H.M., Buluç, A., Williams, S., Yang, C.: Optimizing sparse matrix-multiple vectors multiplication for nuclear configuration interaction calculations. In: International Parallel and Distributed Processing Symposium (IPDPS), pp. 1213–1222 (2014)
4. Arbenz, P., Hetmaniuk, U.L., Lehoucq, R.B., Tuminaro, R.: A comparison of eigensolvers for large-scale 3D modal analysis using AMG-preconditioned iterative methods. Int. J. Numer. Methods Eng. **64**, 204–236 (2005)
5. Avron, H., Toledo, S.: Randomized algorithms for estimating the trace of an implicit symmetric positive semi-definite matrix. J. ACM **58**, 8:1–8:34 (2011)
6. Baker, C.G., Hetmaniuk, U.L., Lehoucq, R.B., Thornquist, H.K.: Anasazi software for the numerical solution of large-scale eigenvalue problems. ACM Trans. Math. Softw. **36**, 1–23 (2009)
7. van Barel, M.: Designing rational filter functions for solving eigenvalue problems by contour integration. Linear Algebra Appl. **502**, 346–365 (2016)
8. Barrett, R., et al.: Templates for the Solution of Linear Systems: Building Blocks for Iterative Methods. SIAM, Philadelphia (1994)

9. Bekas, C., Kokiopoulou, E., Saad, Y.: Polynomial filtered Lanczos iterations with applications in density functional theory. SIAM J. Matrix Anal. Appl. **30**, 397–418 (2008)
10. Bell, R.J., Dean, P.: Atomic vibrations in vitreous silica. Discuss. Faraday Soc. **50**, 55–61 (1970)
11. Berthier, L., Charbonneau, P., Jin, Y., Parisi, G., Seoane, B., Zamponi, F.: Growing timescales and lengthscales characterizing vibrations of amorphous solids. Proc. Nat. Acad. Sci. **113**, 8397–8401 (2016)
12. Derlet, P.M., Maaß, R.: Thermal processing and enthalpy storage of a binary amorphous solid: a molecular dynamics study. J. Mater. Res. **32**, 2668–2679 (2017)
13. Derlet, P.M., Maaß, R.: Local volume as a robust structural measure and its connection to icosahedral content in a model binary amorphous system. Materialia **3**, 97–106 (2018)
14. Derlet, P.M., Maaß, R.: Thermally-activated stress relaxation in a model amorphous solid and the formation of a system-spanning shear event. Acta Mater. **143**, 205–213 (2018)
15. Derlet, P.M., Maaß, R., Löffler, J.F.: The Boson peak of model glass systems and its relation to atomic structure. Eur. Phys. J. B **85**, 1–20 (2012)
16. Fang, H.R., Saad, Y.: A filtered Lanczos procedure for extreme and interior eigenvalue problems. SIAM J. Sci. Comput. **34**, A2220–A2246 (2012)
17. FUJITSU Server Performance Report PRIMERGY RX2540 M4. White paper, version 1.3. Fujitsu Corporation, 17 November 2018 (2018)
18. Galgon, M., et al.: Improved coefficients for polynomial filtering in ESSEX. In: Sakurai, T., Zhang, S.-L., Imamura, T., Yamamoto, Y., Kuramashi, Y., Hoshi, T. (eds.) EPASA 2015. LNCSE, vol. 117, pp. 63–79. Springer, Cham (2017). https://doi.org/10.1007/978-3-319-62426-6_5
19. Hutchinson, M.F.: A stochastic estimator of the trace of the influence matrix for Laplacian smoothing splines. Commun. Stat. Simulat. Comput. **19**, 433–450 (1990)
20. Jay, L.O., Kim, H., Saad, Y., Chelikowsky, J.R.: Electronic structure calculations for plane-wave codes without diagonalization. Comput. Phys. Commun. **118**, 21–30 (1999)
21. Krämer, L., Di Napoli, E., Galgon, M., Lang, B., Bientinesi, P.: Dissecting the FEAST algorithm for generalized eigenproblems. J. Comput. Appl. Math. **244**, 1–9 (2013)
22. Kreutzer, M., Pieper, A., Hager, G., Wellein, G., Alvermann, A., Fehske, H.: Performance engineering of the kernel polynomial method on large-scale CPU-GPU systems. In: International Parallel and Distributed Processing Symposium (IPDPS), pp. 417–426 (2015)
23. Leonforte, F., Boissière, R., Tanguy, A., Wittmer, J.P., Barrat, J.L.: Continuum limit of amorphous elastic bodies. III. Three-dimensional systems. Phys. Rev. B **72**, 224206 (2005)
24. Li, R., Xi, Y., Vecharynski, E., Yang, C., Saad, Y.: A thick-restart Lanczos algorithm with polynomial filtering for Hermitian eigenvalue problems. SIAM J. Sci. Comput. **38**, A2512–A2534 (2016)
25. Liang, Z., Keblinski, P.: Sound attenuation in amorphous silica at frequencies near the boson peak. Phys. Rev. B **93**, 054205 (2016)
26. Lin, L., Saad, Y., Yang, C.: Approximating spectral densities of large matrices. SIAM Rev. **58**, 34–65 (2016)
27. Marruzzo, A., Schirmacher, W., Fratalocchi, A., Ruocco, G.: Heterogeneous shear elasticity of glasses: the origin of the boson peak. Sci. Rep. **3**, 1407 (2013)

28. Monaco, G., Mossa, S.: Anomalous properties of the acoustic excitations in glasses on the mesoscopic length scale. Proc. Nat. Acad. Sci. **106**, 16907–16912 (2009)
29. di Napoli, E., Polizzi, E., Saad, Y.: Efficient estimation of eigenvalue counts in an interval. Numer. Linear Algebra Appl. **23**, 674–692 (2016)
30. Parlett, B.N.: The Symmetric Eigenvalue Problem. Prentice Hall, Upper Saddle River (1980)
31. Pieper, A., et al.: High-performance implementation of Chebyshev filter diagonalization for interior eigenvalue computations. J. Comput. Phys. **325**, 226–243 (2016)
32. Rivlin, T.J.: An Introduction to the Approximation of Functions. Dover, New York (1981)
33. Saad, Y.: Numerical Methods for Large Eigenvalue Problems, 2nd edn. SIAM, Philadelphia (2011)
34. Schaffner, S.: Using Trilinos to solve large scale eigenvalue problems in amorphous materials. Master's thesis, ETH Zurich, Computer Science Department (2015)
35. Schirmacher, W.: The boson peak. Phys. Status Solidi B **250**, 937–943 (2013)
36. Schirmacher, W., Ruocco, G., Scopigno, T.: Acoustic attenuation in glasses and its relation with the boson peak. Phys. Rev. Lett. **98**, 025501 (2007)
37. Schirmacher, W., Scopigno, T., Ruocco, G.: Theory of vibrational anomalies in glasses. J. Non-Cryst. Solids **407**, 133–140 (2015)
38. Schofield, G., Chelikowsky, J.R., Saad, Y.: A spectrum slicing method for the Kohn–Sham problem. Comput. Phys. Commun. **183**, 497–505 (2012)
39. Shintani, H., Tanaka, H.: Universal link between the boson peak and transverse phonons in glass. Nat. Mater. **7**, 870–877 (2008)
40. Silver, R.N., Röder, H.: Calculation of densities of states and spectral functions by Chebyshev recursion and maximum entropy. Phys. Rev. E **56**, 4822–4829 (1997)
41. Silver, R.N., Röder, H., Voter, A.F., Kress, J.D.: Kernel polynomial approximations for densities of states and spectral functions. J. Comput. Phys. **124**, 115–130 (1996)
42. Sleijpen, G.L.G., van den Eshof, J.: On the use of harmonic Ritz pairs in approximating internal eigenpairs. Linear Algebra Appl. **358**, 115–137 (2003)
43. Sleijpen, G.L.G., van der Vorst, H.A.: A Jacobi–Davidson iteration method for linear eigenvalue problems. SIAM J. Matrix Anal. Appl. **17**, 401–425 (1996)
44. Trefethen, L.N.: Approximation Theory and Approximation Practice. SIAM, Philadelphia (2013)
45. The Trilinos Project Home Page. https://trilinos.github.io
46. Wahnström, G.: Molecular-dynamics study of a supercooled 2-component Lennard–Jones system. Phys. Rev. A **44**, 3752–3764 (1991)
47. Weiße, A., Wellein, G., Alvermann, A., Fehske, H.: The kernel polynomial method. Rev. Mod. Phys. **78**, 275–306 (2006)
48. Wu, K., Simon, H.D.: Thick-restart Lanczos method for large symmetric eigenvalue problems. SIAM J. Matrix Anal. Appl. **22**, 602–616 (2000)
49. Xu, N., Wyart, M., Liu, A.J., Nagel, S.R.: Excess vibrational modes and the Boson peak in model glasses. Phys. Rev. Lett. **98**, 175502 (2007)
50. Yamazaki, I., Tadano, H., Sakurai, T., Ikegami, T.: Performance comparison of parallel eigensolvers based on a contour integral method and a Lanczos method. Parallel Comput. **39**, 280–290 (2013)
51. Zhou, Y., Saad, Y., Tiago, M.L., Chelikowsky, J.R.: Self-consistent field calculations using Chebyshev-filtered subspace iteration. J. Comput. Phys. **219**, 172–184 (2006)

Performance Evaluation of Pseudospectral Ultrasound Simulations on a Cluster of Xeon Phi Accelerators

Filip Vaverka[1]([✉]) [ID], Bradley E. Treeby[2] [ID], and Jiri Jaros[1] [ID]

[1] Faculty of Information Technology, Centre of Excellence IT4Innovations, Brno University of Technology, Bozetechova 2, 612 00 Brno, Czech Republic
{ivaverka,jarosjir}@fit.vutbr.cz
[2] Medical Physics and Biomedical Engineering, University College London, London WC1E 6BT, UK
b.treeby@ucl.ac.uk

Abstract. The rapid development of novel procedures in medical ultrasonics, including treatment planning in therapeutic ultrasound and image reconstruction in photoacoustic tomography, leads to increasing demand for large-scale ultrasound simulations. However, routine execution of such simulations using traditional methods, e.g., finite difference time domain, is expensive and often considered intractable due to the computational and memory requirements. The k-space corrected pseudospectral time domain method used by the k-Wave toolbox allows for significant reductions in spatial and temporal grid resolution. These improvements are achieved at the cost of all-to-all communication, which are inherent to the multi-dimensional fast Fourier transforms. To improve data locality, reduce communication and allow efficient use of accelerators, we recently implemented a domain decomposition technique based on a local Fourier basis.

In this paper, we investigate whether it is feasible to run the distributed k-Wave implementation on the Salomon cluster equipped with 864 Intel Xeon Phi (Knight's Corner) accelerators. The results show the immaturity of the KNC platform with issues ranging from limited support of Infiniband and LustreFS in Intel MPI on this platform to poor performance of 3D FFTs achieved by Intel MKL on the KNC architecture. Yet, we show that it is possible to achieve strong and weak scaling comparable to CPU-only platforms albeit with the runtime $1.8\times$ to $4.3\times$ longer. However, the accounting policy for Salomon's accelerators is far more favorable and thus their employment reduces the computational cost significantly.

Keywords: Ultrasound simulations · Local Fourier basis decomposition · Pseudospectral methods · Ultrasound · k-Wave toolbox · Intel Xeon Phi · Knight's Corner · MKL · MPI · OpenMP

© Springer Nature Switzerland AG 2021
T. Kozubek et al. (Eds.): HPCSE 2019, LNCS 12456, pp. 99–115, 2021.
https://doi.org/10.1007/978-3-030-67077-1_6

1   Introduction

There is a growing number of medical applications of ultrasound such as pho-toacoustic imaging [1], neurostimulation [28] and high intensity focused ultra-sound (HIFU) cancer treatment [5,18]. These applications require fast, accurate and versatile ultrasound propagation models in tissue-like materials at various stages such as planning or post-processing. Typically, these requirements can be met with models based on the generalized Westervelt equation [22], which allows for modeling nonlinear ultrasound wave propagation through heteroge-neous medium with a power law absorption. Due to this demand, several ultra-sound modeling packages for medical applications have been released along with our k-Wave toolkit, see [9] for a recent review. The majority of those pack-ages employ either finite-difference time-domain (FDTD) method, pseudospec-tral time-domain (PSTD) method or a variant of operator-splitting methods. The k-Wave toolbox is among a few based on the k-space pseudo-spectral time-domain (KSTD) method.

FDTD methods scale well on large parallel systems using a straightforward domain decomposition and halo exchange over the nearest neighbors [31]. How-ever, most FDTD schemes require between 8 to 10 grid points per wavelength to achieve sufficient accuracy and even more to manage dispersion in cases where propagation over a large number of wavelengths has to be modeled [17]. This makes many realistic ultrasound simulations intractable due to high mem-ory requirements. The PSTD methods can theoretically approach the Nyquist limit of 2 grid points per wavelength and thus significantly reduce the memory requirements. The KSTD method improves on the PSTD method by using a semi-analytical time-stepping schemes [25] which complements excellent spatial properties with a larger time step size. The main drawback of PSTD and espe-cially KSTD methods is the introduction of a global trigonometric basis and the use of the fast Fourier transform (FFT) to implement gradient operators. The KSTD method requires, due to the k-space correction, 3D FFTs which inher-ently limit the scalability of this method on parallel distributed, and especially on accelerated systems [13]. Although a lot of work on efficient distributed FFTs has been carried out (FFTW [7], Hybrid FFTW [20], P3DFFT [21], PFFT [23], AccFFT [8], multi-GPU CUDA FFT [19] or FFT-ECP [26]), the computation time is still often determined by the communication between subdomains, which in many cases prevents the use of accelerators such as GPUs or Intel Xeon Phis.

This paper investigates the possibility of deploying a distributed implementa-tion of the KSTD method implemented in the k-Wave toolbox [14] on a large clus-ter of Intel Xeon Phi accelerators. The method combines advantages of FDTD and KSTD methods by replacing global Fourier basis with a set of local ones [12], thus achieving communication complexity of an FDTD method while maintain-ing many properties of a KSTD method. The following section briefly describes the principle of the local Fourier basis decomposition. Next, the architecture of the accelerated cluster is described in Sect. 3. After that, the implementation is briefly explained in Sect. 4 and achieved scaling results are presented in Sect. 5.

Finally, Sect. 6 investigates performance of the accelerators and the last section draws conclusions on usability of the platform.

2   Local Fourier Basis Domain Decomposition

The numerical model of the nonlinear wave propagation in heterogeneous absorbing medium investigated in this paper is based on the governing equations derived by Treeby [27] written as three-coupled first-order partial differential equations:

$$\frac{\partial \mathbf{u}}{\partial t} = -\frac{1}{\rho_0}\nabla p + \mathbf{F} \ , \qquad\qquad \text{(momentum conservation)}$$

$$\frac{\partial \rho}{\partial t} = -\rho_0 \nabla \cdot \mathbf{u} - \mathbf{u} \cdot \nabla \rho_0 - 2\rho \nabla \cdot \mathbf{u} + M \ , \qquad \text{(mass conservation)} \quad (1)$$

$$p = c_0^2 \left(\rho + \mathbf{d} \cdot \nabla \rho_0 + \frac{B}{2A}\frac{\rho^2}{\rho_0} - \mathrm{L}\rho \right). \qquad \text{(equation of state)}$$

Here $\mathbf{u}$ is the acoustic particle velocity, $\mathbf{d}$ is the acoustic particle displacement, p is the acoustic pressure, ρ is the acoustic density, ρ_0 is the ambient (or equilibrium) density, c_0 is the isotropic sound speed, B/A is the nonlinearity parameter, and operator L captures power law absorption and dispersion. Two linear source terms are also included, where $\mathbf{F}$ is a force source term, and M is a mass source.

In the k-Wave toolbox, the model (1) would normally be directly discretized using the KSTD method (see [27]). However, to alleviate the global communication overhead, we instead divide the simulation domain into a set of cuboid subdomains, each of which supported by its own local Fourier basis [12] (LFB for short). The neighboring subdomains are coupled by overlapping each other and the overlaps are also used to restore their local periodicity [2], see Fig. 1.

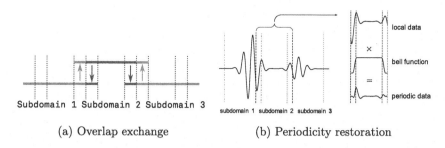

(a) Overlap exchange   (b) Periodicity restoration

Fig. 1. The principle of local Fourier basis domain decomposition shown for one spatial dimension. (a) The local subdomain is padded with a few grid points (overlap) from both neighboring subdomains which are periodically exchanged. (b) After the overlap exchange, each local domain is multiplied by a bell function to restore periodicity.

The computation itself consists of an iterative algorithm running over a given number of time steps. Each time step is composed of a sequence of element-wise operations, overlap exchanges and local 3D FFTs, see Fig. 2. The majority of the computation time is usually spent on 3D FFTs or overlap exchanges. The restriction of the Fourier basis to the local subdomain has naturally a negative impact on the accuracy of the LFB method. The amount of accuracy loss depends on the overlap size and the properties of the bell function used [3, 14, 29].

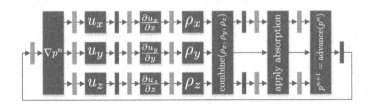

Fig. 2. Simplified computation loop governed by Eq. (1). Blue blocks denote element-wise operations, yellow 3D FFTs, and orange overlap exchanges. (Color figure online)

3   Target Architecture

The target architecture investigated in this paper is a typical Intel/Infiniband cluster accelerated with Intel Xeon Phi cards of the Knight's Corner architecture [15]. The experiments were conducted on the Salomon supercomputer at the IT4Innovations national supercomputing center in Ostrava, Czech Republic[1].

Salomon consists of 1008 compute nodes, 432 of which are accelerated by Intel Xeon Phi 7120P accelerators. The architecture of the Salomon's accelerated part is shown in Fig. 3. Every node consists of a dual socket motherboard populated with two Intel Xeon E5-2680v3 (Haswell) processors accompanied with 128 GB of RAM. The nodes also integrate a pair of accelerators connected to individual processor sockets via the PCI-Express 2.0 x16 interface. The communication between processors is handled by the Intel QPI interface.

The nodes are interconnected by a 7D enhanced hypercube running on the 56 Gbit/s FDR Infiniband technology. The accelerated nodes occupy a subset of this topology constituting a 6D hypercube. Every node contains a single Infiniband network interface (NIC) connected via PCI-Express 3.0 to the first socket, and a service 1 Gbit/s Ethernet interface connected to the same socket. Both accelerators are capable of directly accessing the Infiniband NIC by means of Remote Direct Memory Access (RDMA).

---

[1] https://docs.it4i.cz/salomon/hardware-overview/.

A single Intel Xeon Phi 7120P accelerator packs 61 P54C in-order cores extended by 4-wide simultaneous multithreading (SMT) and a 512-bit wide vector processing unit (VPU). The Xeon Phi cores are supported by 30.5 MB of L2 cache evenly distributed over individual cores and interconnected via a ring bus. The memory subsystem consists of 4 memory controllers managing in total 16 GB of GDDR5. The theoretical performance and memory bandwidth of a single accelerator is over 2 TFLOP/s in single precision and 352 GB/s, respectively. A single accelerator can theoretically provide a speedup of 4× for compute bound, and 5× for memory bandwidth bound applications over a single twelve core Haswell processor. The total compute power of the accelerated part of the cluster reaches one PFLOP/s.

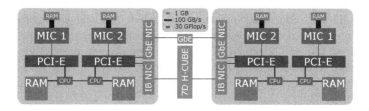

Fig. 3. The architecture of the Salomon accelerated nodes and interconnection. The size of the rectangles representing individual components is proportional to their performance, bandwidth or capacity.

4   Implementation

4.1   Execution Mode

The Intel Xeon Phi offers three different modes of execution: (1) The offload mode is an analogy to the GPGPU approach where the accelerator is controlled by the CPU and used only for the compute intensive tasks. (2) The native mode uses the accelerator as an isolated compute node with shared memory. (3) The cluster mode allows accelerators to be connected using the message passing interface (MPI) and run distributed jobs over many accelerators. In this mode, the CPUs can be used either to only run the operating system, or join the distributed job as additional workers, although with reduced compute power.

The proposed implementation uses Intel Xeon Phis in the native mode, which allows for direct access to NICs through the RDMA mechanism and thus should achieve the best performance in MIC-to-MIC communication. Overall the PSTD and KSTD methods tend to be memory bound as they exhibit relatively low arithmetic intensity on the order of $O(\log n)$. Therefore, the k-Wave toolbox favors architectures with high memory bandwidth and fast access to the network.

The code is logically structured into one MPI process per subdomain which runs on a single accelerator (or CPU socket). The work local to each subdomain is distributed across cores by means of OpenMP and OpenMP SIMD constructs. Since realistic simulations do not require double precision, only single precision floating point operations are used. This yields higher performance and saves valuable memory bandwidth. The k-Wave LFB code boils down to a mix of element-wise operations on 3D real or complex matrices, 3D fast Fourier transforms and overlap exchanges. The following sections briefly describe most important operations, however, more details can be found in [29].

4.2 Fast Fourier Transforms

The most computationally expensive part of the simulation loop consists of a number of 3D fast Fourier transforms calculated over the local subdomains. Depending on which medium parameters are heterogeneous, the number of FFTs varies between ten and fourteen. Their actual implementation relies on third party libraries compatible with the FFTW interface [7], in this case the Intel MKL library [11] which showed better results than FFTW on the Intel Xeon Phi architecture [30].

The simulation code mostly uses out-of-place real-to-complex and complex-to-real transforms which reduces both the spatial and temporal complexity of the FFT by a factor of two [24]. However, the implementation of the out-of-place C2R transforms in the MKL library has proved to be very inefficient on the Intel Xeon Phi Knight's Corner. Hence, the C2R transforms are performed in-place using a temporary matrix and the results consequently copied to the destination matrix. The performance characteristics of the 3D FFT implementation are further analyzed in Sect. 6.2.

4.3 Overlap Exchanges

The gradient or derivative calculation on a subdomain can be performed only after gathering the most recent data from all its neighbors. This is accommodated by exchanging the overlap regions between neighboring subdomains before every such operation, see the orange bars in Fig. 2. The size of these transfers range from N_d^3 to $N_x \times N_y \times N_d$ grid points, where N_d is the overlap size and $N_{\{x,y,z\}}$ the subdomain edge length.

Since the overlaps have to contain the most recent data, it is difficult to properly hide the communication by overlapping it with useful computation. However, our implementation structurally allows to hide up to 50% of the communication by exploiting stages where multiple arrays have to be updated at the same time. This is achieved by a combination of persistent communications and non-blocking calls provided by MPI, see Listing 1.

```
1   // Initialization stage
2   for (auto &m: /* Velocity matrices U_x, U_y, U_z */) {
3     for (auto &n: m.getNeighbors()) {
4       MPI_Send_init(n.data, n.size, n.otherRank, /* ... */);
5       MPI_Recv_init(n.data, n.size, n.otherRank, /* ... */);
6     }
7   }
8
9   // Main simulation loop stage
10  for (auto &m: /* Velocity matrices U_x, U_y, U_z */)
11    MPI_Startall(m.getRequests().size(), m.getRequests().data());
12
13  for (auto &m: /* Velocity matrices U_x, U_y, U_z */) {
14    // Partially overlapped communication
15    MPI_Waitall(m.getRequests().size(), m.getRequests().data(),
16               /* ... */);
17    Compute_Forward_FFT_3D(m);
18  }
```

Listing 1: The principle of the communication overlap during the velocity gradient calculation. A set of persistent communications are created during the initialization stage (Line 1–7). The overlap exchange for multiple matrices is started in the simulation loop (Line 9–11). As soon as the communication for a given matrix has finished, the computation starts. The other transfers can be still in flight (Line 13–17).

4.4 Parallel Input and Output

The simulation of nonlinear ultrasound wave propagation in large realistic domains naturally requires a fast and scalable input-output subsystem. Not only may typical medium, source and input signal descriptions occupy tens to hundreds of GBs, the simulation outputs in the form of pressure and velocity time series can easily spread over a few TBs [16].

The parallel I/O subsystem of the k-Wave toolbox is based on the Parallel HDF5 library [6] supported by the MPI-IO back-end [4]. In combination with the parallel LustreFS file system, very good throughput reaching up to 15 GB/s has been achieved on several clusters [10].

Although the parallel I/O is not intended to be deeply investigated in this paper, there is a key observation that needs to be mentioned because of it being a significant source of issues on the accelerated part of the Salomon supercomputer. The obstacle is the lack of LustreFS support in the Xeon Phi implementation of the Intel MPI library. This makes the use of the distributed scratch file system impossible and forces us to use an NFS mounted home file system, not primarily intended for parallel I/O. As a consequence, the amount of data being collected was severely limited to avoid any inference in the experiments.

5   Scaling Results

5.1   Overview

The main objective of this section is to investigate the performance and scaling properties of the cluster of Intel Xeon Phi accelerators in simulating ultrasound wave propagation. The secondary objective is the evaluation of the software support and the ease of use of the whole platform.

The experiments were conduced on various numbers of Salomon's accelerated nodes ranging from 1 up to 256 (512 accelerators). Since Salomon's hypercube interconnection topology has some blocking factor, there is a measurable variation between the job instances stemming from the different placement of the MPI processes on currently available nodes. Although the job placement can be restricted manually, it significantly prolongs the waiting time in the queues and thus was not used. Instead, particular benchmark runs of the same type were packed into a single large job to maintain fair conditions. Different job allocations were used for benchmarking subdomains placed over CPUs and accelerators. Therefore, a small variation between these benchmarks may be observed, however, it is considered insignificant from the perspective of the overall scaling trends and even the absolute performance.

Every benchmark run consisted of 100 time steps of the simulation loop summarized in Fig. 2. This number is deemed sufficient to hide any cache and communication warm-up effects. All experiments, if not stated otherwise, were performed with the most typical overlap size of 16 grid points.

The following subsections first focus on obtaining the performance and scaling results for the largest possible simulation domains while trying to work around the stability and HW/SW support issues. However, some issues are related to high numbers of nodes used for large simulation domain sizes. It was not possible to resolve these issues even after consultation with the Salomon support and the vendor. In the second subsection, we thus limited the simulation domain size and the number of nodes used to mimic an ideal situation. Finally, the most significant stages of the KSTD solver are analyzed and discussed.

5.2   Performance Scaling on Large Domains

Driven by the practical demands from industry and medical physics, we first focus on the performance scaling of large simulation domains spread over many accelerators. The simulation domain sizes of interest are expanded from $256 \times 256 \times 256$ (2^{24}) to $2048 \times 2048 \times 2048$ (2^{33}) grid points by sequentially doubling the dimension sizes starting from the least significant one. The domains are partitioned into a number of subdomains growing from 1 to 256. The numbers of subdomains for particular domain sizes are further restricted by the size of the smallest meaningful subdomain (64^3) and the largest possible subdomain ($256 \times 256 \times 512$) that can fit within memory, excluding the overlaps. Particular subdomains are assigned either to a single accelerator or a single CPU socket. This allows us to mutually compare performance scaling of both architectures.

Figures 4 and 5 show the strong and weak scaling achieved by both architectures. Although the whole range of overlap sizes between 2 to 32 grid points
was investigated, only the most common overlap size of 16 is presented for the
sake of brevity. The scaling of small overlap sizes generally does better due to a
higher degree of communication overlapping. For bigger overlap sizes, the absolute execution time is more influenced by the communication time elaborated in
Fig. 8 and the strong scaling curves appear flatter.

A brief glance at Fig. 4 reveals a significant disparity between the performance
of CPUs and accelerators. The dramatic slowdown on the accelerator side is
caused by the communication layer, more specifically the Intel MPI Infiniband
backend DAPL [15], which becomes unstable for more than 32 accelerators in a
single job. This problem was discussed in detail with the cluster support team
and the vendor specialists but has not been resolved yet. The only solution that
proved to be stable was to replace the 56 Gbit Infiniband by a 1 Gbit Ethernet
interface. This typically leads to a 4.3× slower execution compared to CPUs.
Nevertheless the code maintains reasonable strong scaling factors of 1.45 (average
speedup achieved by doubling the computational resources) comparable with
the execution on CPUs. The only reason why the performance slump is not even
deeper is the limited performance of the FFTs on the accelerators, and therefore,
better opportunity for communication overlapping.

Examining further the CPU and accelerator scaling plots, there are very few
anomalies in the scaling curves. The most significant are apparent for very small
numbers of subdomains when using accelerators, or for tiny subdomains when
using CPUs. Putting these results into context of the global Fourier basis decom-
positions (GFB) presented in [13], the LFB implementation on CPUs shows its
superiority with typical and peak speedup of 2 and 6, respectively.

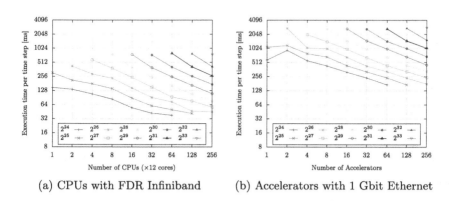

(a) CPUs with FDR Infiniband   (b) Accelerators with 1 Gbit Ethernet

Fig. 4. Strong scaling evaluation on large domains having between 2^{24} and 2^{33} grid
points with an overlap size of 16 grid points collected on CPUs and accelerators. Since
the Infiniband interface is not stable for more than 32 accelerators, 1 Gbit Ethernet is
used instead.

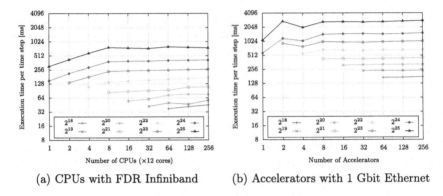

(a) CPUs with FDR Infiniband   (b) Accelerators with 1 Gbit Ethernet

Fig. 5. Weak scaling evaluation on large subdomains with overlap size of 16 grid points collected on CPUs and accelerators. The size of the subdomains ranges between 2^{18} and 2^{25} grid points.

Figure 5 shows the weak scaling achieved on the CPUs and accelerators. Each of the plotted series corresponds to a constant subdomain size from the investigated range between 64^3 and $256 \times 256 \times 512$ grid points. At first glance, poor weak scaling is observed when the simulation domain is split into less than 8 subdomains. This is due to the growing rank of the domain decomposition and the number of neighbors. Once a full 3D decomposition is reached, the scaling curves remain almost flat in-line with almost perfect scaling.

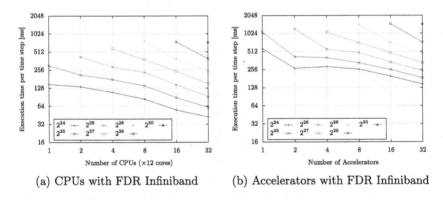

(a) CPUs with FDR Infiniband   (b) Accelerators with FDR Infiniband

Fig. 6. Strong scaling evaluation on small domains (2^{24} to 2^{30} grid points) with an overlap size of 16 grid points collected on CPUs and accelerators, both supported by FDR Infiniband.

5.3 Performance Scaling on Small Domains

In order to better quantify the impact of the misbehaving Infiniband on large jobs, the previous experiments were repeated with reduced domain sizes (up to

1024^3) and a reduced number of accelerators (up to 32). Figure 6 shows the improved strong scaling on accelerators when using Infiniband and compares it again with the CPU baseline. Not only does Infiniband reduce the overall execution time by a factor of 1.9 with respect to Ethernet, it also improves the scaling factors from 1.45 to 1.52. The reduction of the communication overhead has also a positive impact on weak scaling presented in Fig. 7. Here, the penalty caused by increasing the rank of the decomposition is greatly reduced and very good scaling is achieved even when going from one to two subdomains. Nevertheless, the final conclusion is that a cluster of Intel Xeon Phi accelerators is significantly slower than a comparable cluster of CPUs, even though the theoretical parameters of the architecture promise the direct opposite.

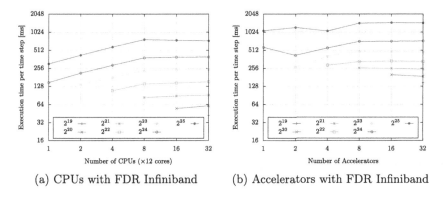

(a) CPUs with FDR Infiniband   (b) Accelerators with FDR Infiniband

Fig. 7. Weak scaling evaluation on small domains (subdomains of 2^{19} to 2^{25} grid points) with an overlap size of 16 grid points collected on CPUs and accelerators, both supported by FDR Infiniband.

5.4   Simulation Time Breakdown

Figure 8 shows the execution time breakdown for domains of various sizes partitioned into 32 ($2 \times 4 \times 4$) and 256 ($4 \times 8 \times 8$) subdomains with an overlaps size of 16 grid points. The slower interconnection of the accelerators becomes immediately evident from Fig. 8b which shows that this usually comprises more than 60% of the compute time. By comparing the communication time on the accelerators with CPUs we can find a massive 3× to 12× deterioration which increases proportionally with the simulation size. On the other hand, the calculation time remains favorable for medium sized subdomains. For small ones, there is not enough work for all 120 threads used. On contrary the L2 cache is exceeded for subdomains bigger than 256^3 grid points, which leads to a significant drop in the FFT performance (see Fig. 8a at 2^{29} and 2^{30}). In conclusion, the computation can only be 1.8× slower than a single CPU socket for favorable

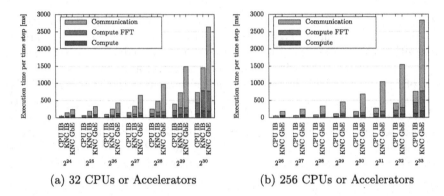

(a) 32 CPUs or Accelerators   (b) 256 CPUs or Accelerators

Fig. 8. The execution time breakdown for a time step of the simulation loop collected on 32 and 256 CPUs (CPU) or accelerators (KNC) for different domain sizes with an overlap size of 16 grid points. Results for both the Infiniband (IB) and the 1 Gbit Ethernet (GbE) interconnects are shown.

subdomain sizes while it can worsen to more than $16\times$ slower for large subdomains. Both interconnect and FFT computation issues are further investigated in Sect. 6.

6   Platform Investigation

6.1   Overview

As mentioned earlier, the practical use of a cluster of Intel Xeon Phi accelerators suffers from many issues and immature libraries. Apart from the missing support for the LustreFS file system and the instability of the Intel MPI communication back-end for the infiniband interface, there are two other major issues. The first is the performance of 3D FFTs and the second one is the communication bandwidth. Both are further investigated in this section.

6.2   Performance of 3D FFTs on Intel Xeon Phi

Since the performance breakdown presented in Fig. 8 revealed relatively poor performance of the FFTs running on the accelerators compared to a single CPU, a deeper performance investigation was carried out. The particular routines of interest were the real-to-complex (R2C) and complex-to-real (C2R) 3D FFTs with the conjugate-even storage within an interleaved complex array implemented in the Intel MKL library. Both in-place and out-of-place transforms were evaluated.

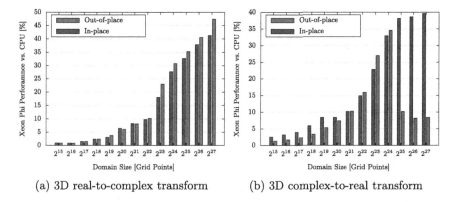

(a) 3D real-to-complex transform   (b) 3D complex-to-real transform

Fig. 9. Performance of a single accelerator vs. a single CPU while running in-place and out-of place forward and inverse 3D FFTs implemented by Intel MKL.

Figure 9 shows the relative performance of a single accelerator executing 120 threads versus a single CPU executing 12 threads. The thread counts were chosen according to the best performance achieved by each architecture. The measured performance is plotted for several domain sizes growing from 32^3 to 512^3 grid points.

Although the theoretical values suggest an accelerator should be four times faster than a single CPU, our FFT benchmarks revealed a different picture. For small domains, the performance of the accelerators is dismal, reaching less than 10% compared to a single CPU. This is very likely caused by poor workload distribution among cores and cache coherency issues such as false sharing. However, even a 3D transform over a 64^3 domain requires the execution of 3×4096 independent 1D FFTs, which appears to be enough to employ 120 threads evenly. Moreover, as the size of each 1D FFT is only 64 elements (256 B, 1 MB in total), the capacity of L2 cannot be a true bottleneck. That being said, the 1D transforms are unlikely to cause the troubles. The other sources of performance issues are the multi-dimensional matrix transpositions and thread synchronization between particular phases of the 3D FFT algorithm.

With an increasing transform size, the performance of the accelerator increases, reaching 50% of the CPU performance in the best case. Furthermore, the performance of the forward transforms is slightly better than the performance of the inverse one. Much more unexpected behavior is observed for the out-of-place transforms on domains larger than 256^3 grid points. Beyond this size, the relative performance of the accelerator falls below 10%. At this size, the data cannot fit within L2 cache any more. Considering the only difference between the forward and inverse FFT being the sign in the exponent, this behavior must be a hidden bug inside the complex-to-real out-of-place FFTs in MKL. This statement is supported by the measurements provided by Intel VTune performance counters. At a size of $256 \times 256 \times 512$ grid points, there is a 60-fold increase in the L1 data cache misses, most of which end up as L2 data cache memory fills. These misses seem to be caused by the Read for Ownership (RFO) operations executed before memory writes recorded as `L2_DATA_WRITE_MISS_MEM_FILL` events, a typical sign of false sharing. The issue is related to the FFT execution plan and the memory hierarchy as out-of-place real-to-complex transforms do not exhibit similar behavior.

6.3   Performance of Intel MPI on Intel Xeon Phi

The second issue afflicting the performance is the interconnection. Even when using the Infiniband interface, the accelerators do not achieve comparable communication times to CPUs as shown in Fig. 8a.

Figure 10 shows the bidirectional bandwidth and latency in CPU-to-CPU and Phi-to-Phi communication over the Infiniband and Ethernet interconnects measured by the OSU Micro-Benchmark suite[2]. The difference is obvious. Not only is there an order of magnitude lower bandwidth for small messages when the infiniband is used between accelerators, the loss is not caught up even for medium sized messages. There appears to be an improvement for 4 MB messages, however, this is helpful only for the biggest subdomains with an overlap size of 32 grid points. The communication latency copies the same trend, being typically 5× to 10× longer. The explanation of this behavior can be found in the combination of the PCI-Express 2.0 (about 16 GB/s bidirectional), additional hop between PCI-Express 2.0 and PCI-Express 3.0 where the Infiniband card is connected to, and low attainable memory bandwidth of a single Intel Xeon Phi core of (only about 2.65 GB/s) yielding 4.1 GB/s for bidirectional communication.

The bandwidth of the Ethernet interface is, as expected, limited by the network interface for big messages.

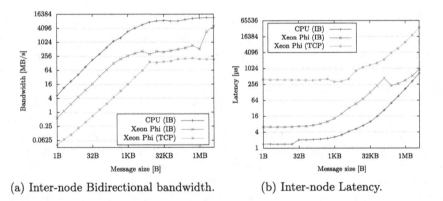

(a) Inter-node Bidirectional bandwidth.   (b) Inter-node Latency.

Fig. 10. Comparison of the communication bandwidth and latency for processors with Infiniband, and accelerators with Infiniband and TCP over 1 Gbit Ethernet.

---

[2] http://mvapich.cse.ohio-state.edu/benchmarks/.

7   Conclusion

The goal of this paper was to investigate the performance scaling and suitability of accelerated clusters based on Intel Xeon Phis for large simulations of ultrasound wave propagation in biologically relevant materials, and compare these results with a common CPU cluster.

In order to keep low requirements on the spatial and temporal resolution, the k-space corrected pseudospectral method from the k-Wave toolbox was used [27]. The communication overhead is reduced by the local Fourier basis decomposition [14] and the communication is overlapped with the computation in more than 50% of cases.

The performance was measured on a set of domain sizes starting from tiny ones and growing to the edge of practical feasibility. During testing, many problems were encountered. The most significant one is the Infiniband network instability caused by the DAPL back-end in the Intel MPI. This bug made the use of the native Infiniband impossible for jobs spreading over more than 32 accelerators. Despite careful investigation and collaboration with the Salomon support, this issue has not been resolved yet. The only solution to get large simulations to work was to use a service Ethernet network. This naturally has a large impact on performance. Regardless, the scaling was similar to the underlying cluster of CPUs, which is caused by another issue related to the FFT performance, and the absolute execution time is typically 4.3× longer. This is far from the expected performance inferred from the theoretical parameters. When the size of the simulation is limited to fit within 32 accelerators, the Infiniband interconnection remains stable. This yields much better performance, but still almost 1.9× lower than CPUs.

The second big issue is the performance of 3D FFTs, which in some configurations reaches only a fraction of the CPU performance. This is very likely caused by a bug in the MKL library related to false sharing during the matrix transposition between particular 1D FFTs. The best performance obtained reached about 50% of the CPU performance.

Despite all the mentioned troubles, we are still positive about the code deployment on the accelerated cluster. The main motivation is the accounting policy on Salomon, where the use of accelerators is now for free. This allows us to run large batches of independent simulations. In the future, we would also like to implement a load-balancing algorithm that would allow us to use both the CPUs and the accelerators in a single simulation.

Acknowledgement. This work was supported by The Ministry of Education, Youth and Sports from the National Programme of Sustainability (NPU II) project "IT4Innovations excellence in science - LQ1602" and by the IT4Innovations infrastructure which is supported from the Large Infrastructures for Research, Experimental Development and Innovations project "IT4Innovations National Supercomputing Center - LM2015070". This project has received funding from the European Union's Horizon 2020 research and innovation programme H2020 ICT 2016–2017 under grant agreement No. 732411 and is an initiative of the Photonics Public Private Partnership.

This work was further supported in part by the Engineering and Physical Sciences Research Council (EPRSC), UK, grant numbers EP/L020262/1 and EP/S026371/1.

References

1. Beard, P.: Biomedical photoacoustic imaging. Interf. Focus **1**(4), 602–631 (2011)
2. Boyd, J.P.: A comparison of numerical algorithms for Fourier extension of the first, second, and third kinds. J. Comput. Phys. **178**(1), 118–160 (2002)
3. Boyd, J.P.: Asymptotic Fourier coefficients for a $C\infty$ bell (Smoothed-"Top-Hat") & the Fourier extension problem. J. Sci. Comput. **29**(1), 1–24 (2006)
4. Coloma, K., et al.: A new flexible MPI collective I/O implementation. In: 2006 IEEE International Conference on Cluster Computing, pp. 1–10. IEEE (2006)
5. Dubinsky, T.J., Cuevas, C., Dighe, M.K., Kolokythas, O., Joo, H.H.: High-intensity focused ultrasound: current potential and oncologic applications. Am. J. Roentgenol. **190**(1), 191–199 (2008)
6. Folk, M., Heber, G., Koziol, Q., Pourmal, E., Robinson, D.: An overview of the HDF5 technology suite and its applications. In: Proceedings of the EDBT/ICDT 2011 Workshop on Array Databases - AD 2011 (2011)
7. Frigo, M., Johnson, S.G.: The design and implementation of FFTW3. Proc. IEEE **93**(2), 216–231 (2005)
8. Gholami, A., Hill, J., Malhotra, D., Biros, G.: AccFFT: a library for distributed-memory FFT on CPU and GPU architectures (2016)
9. Gu, J., Jing, Y.: Modeling of wave propagation for medical ultrasound: a review. IEEE Trans. Ultrason. Ferroelectr. Freq. Control **62**(11), 1979–1992 (2015)
10. Howison, M., Koziol, Q., Knaak, D., Mainzer, J., Shalf, J.: Tuning HDF5 for Lustre file systems. In: Proceedings of the Workshop on Interfaces and Abstractions for Scientific Data Storage 5, IASDS 2010 (2012)
11. Intel Corporation: Math Kernel Library 11.3 Developer Reference. Intel Corporation (2015)
12. Israeli, M., Vozovoi, L., Averbuch, A.: Spectral multidomain technique with local Fourier basis. J. Sci. Comput. **8**(2), 135–149 (1993)
13. Jaros, J., Rendell, A.P., Treeby, B.E.: Full-wave nonlinear ultrasound simulation on distributed clusters with applications in high-intensity focused ultrasound. Int. J. High Perform. Comput. Appl. **30**(2), 137–155 (2016)
14. Jaros, J., Vaverka, F., Treeby, B.E.: Spectral domain decomposition using local Fourier basis: application to ultrasound simulation on a cluster of GPUs. Supercomput. Front. Innov. **3**(3), 40–55 (2016)
15. Jeffers, J., Reinders, J.: Intel Xeon Phi Coprocessor High Performance Programming. Elsevier Inc., Waltham (2013). No. 1
16. Klepárník, P., Bařina, D., Zemčík, P., Jaroš, J.: Efficient low-resource compression of HIFU data. Information **9**(7), 1–14 (2018). https://doi.org/10.3390/info9070155. https://www.fit.vut.cz/research/publication/11764
17. Mast, T., Souriau, L., Liu, D.L., Tabei, M., Nachman, A., Waag, R.: A k-space method for large-scale models of wave propagation in tissue. IEEE Trans. Ultrason. Ferroelectr. Freq. Control **48**(2), 341–354 (2001)
18. Meairs, S., Alonso, A.: Ultrasound, microbubbles and the blood-brain barrier. Prog. Biophys. Mol. Biol. **93**(1–3), 354–362 (2007)

19. Nandapalan, N., Jaros, J., Treeby, B.E., Rendell, A.P.: Implementation of 3D FFTs across multiple GPUs in shared memory environments. In: Proceedings of the Thirteenth International Conference on Parallel and Distributed Computing, Applications and Technologies, pp. 167–172 (2012)

20. Nikl, V., Jaros, J.: Parallelisation of the 3D fast Fourier transform using the hybrid OpenMP/MPI decomposition. In: Hliněný, P., et al. (eds.) MEMICS 2014. LNCS, vol. 8934, pp. 100–112. Springer, Cham (2014). https://doi.org/10.1007/978-3-319-14896-0_9

21. Pekurovsky, D.: P3DFFT: a framework for parallel computations of Fourier transforms in three dimensions (2012)

22. Pinton, G.F., Dahl, J., Rosenzweig, S., Trahey, G.E.: A heterogeneous nonlinear attenuating full-wave model of ultrasound. IEEE Trans. Ultrason. Ferroelectr. Freq. Control 56(3), 474–488 (2009)

23. Pippig, M.: PFFT-an extension of FFTW to massively parallel architectures. SIAM J. Sci. Comput. 35(3), 213–236 (2013)

24. Sorensen, H., Jones, D., Heideman, M., Burrus, C.: Real-valued fast Fourier transform algorithms. IEEE Trans. Acoust. Speech Signal Process. 35(6), 849–863 (1987)

25. Tabei, M., Mast, T.D., Waag, R.C.: A k-space method for coupled first-order acoustic propagation equations. J. Acoust. Soc. Am. 111(1 Pt 1), 53–63 (2002)

26. Tomov, S., Haidar, A., Ayala, A., Schultz, D., Dongarra, J.: FFT-ECP fast Fourier transform, 01 2019 (2019)

27. Treeby, B.E., Jaros, J., Rendell, A.P., Cox, B.T.: Modeling nonlinear ultrasound propagation in heterogeneous media with power law absorption using a k-space pseudospectral method. J. Acoust. Soc. Am. 131(6), 4324–4336 (2012)

28. Tufail, Y., Yoshihiro, A., Pati, S., Li, M.M., Tyler, W.J.: Ultrasonic neuromodulation by brain stimulation with transcranial ultrasound. Nat. Protoc. 6(9), 1453–1470 (2011)

29. Vaverka, F., Treeby, B.E., Jaros, J.: Evaluation of the suitability of Intel Xeon Phi clusters for the simulation of ultrasound wave propagation using pseudospectral methods. In: Rodrigues, J.M.F., et al. (eds.) ICCS 2019. LNCS, vol. 11538, pp. 577–590. Springer, Cham (2019). https://doi.org/10.1007/978-3-030-22744-9_45

30. Wang, E., et al.: High-Performance Computing on the Intel® Xeon Phi™. Springer, Cham (2014). https://doi.org/10.1007/978-3-319-06486-4

31. Yu, W., Mittra, R., Su, T., Liu, Y., Yang, X.: Parallel Finite-Difference Time-Domain Method. Artech House, Inc., Norwood (2006)

Estimation of Execution Parameters
for k-Wave Simulations

Marta Jaros[1]($\boxtimes$)(iD), Tomas Sasak[1], Bradley E. Treeby[2](iD), and Jiri Jaros[1](iD)

[1] Faculty of Information Technology, Centre of Excellence IT4Innovations,
Brno University of Technology, Bozetechova 2, 612 66 Brno, Czech Republic
{martajaros,jarosjir}@fit.vutbr.cz, xsasak01@stud.fit.vutbr.cz
[2] Medical Physics and Biomedical Engineering, Biomedical Ultrasound Group,
University College London, Malet Place Engineering Building, London
WC1E 6BT, UK
b.treeby@ucl.ac.uk

Abstract. Estimation of execution parameters takes centre stage in automatic offloading of complex biomedical workflows to cloud and high performance facilities. Since ordinary users have no or very limited knowledge of the performance characteristics of particular tasks in the workflow, the scheduling system has to have the capabilities to select appropriate amount of compute resources, e.g., compute nodes, GPUs, or processor cores and estimate the execution time and cost.

The presented approach considers a fixed set of executables that can be used to create custom workflows, and collects performance data of successfully computed tasks. Since the workflows may differ in the structure and size of the input data, the execution parameters can only be obtained by searching the performance database and interpolating between similar tasks. This paper shows it is possible to predict the execution time and cost with a high confidence. If the task parameters are found in the performance database, the mean interpolation error stays below 2.29%. If only similar tasks are found, the mean interpolation error may grow up to 15%. Nevertheless, this is still an acceptable error since the cluster performance may vary on order of percent as well.

Keywords: Workflow management system · Performance data collection · Interpolation · Job scheduling · HPC as a service

1 Introduction

Computation of complex scientific applications may no longer be satisfied by personal computers and small servers manually operated by highly experienced users. First, the extent of data being processed and the computational requirements highly exceed the capacity of such machines. Increasing number of applications is thus moving to the cluster or cloud environments. Second, scientific applications often feature a very complex processing workflows consisting of many particular tasks employing different computer codes, and complex data

© Springer Nature Switzerland AG 2021
T. Kozubek et al. (Eds.): HPCSE 2019, LNCS 12456, pp. 116–134, 2021.
https://doi.org/10.1007/978-3-030-67077-1_7

dependencies. Third, scheduling, execution and monitoring of such workflows require automated tools to remove the burden from the experienced users, enable ordinary users to routinely execute their applications, and increase the through-put of the computing facilities.

To face these challenges, the scientific and software development communities have adopted the workflow paradigm to describe the processing flow. The most common formalism used is the weighted directed acyclic graph (DAG) defining computational tasks by the nodes, and the dependencies and data movements by the edges. The weights in the nodes describe the computational requirements while the weights on the edges denote the amount of data being transferred between tasks [15].

In order to automate workflow execution, several workflow management systems (WMSs) have been developed and used within the scientific community. The most popular tools such as Pegasus [2,3], Globus [4] or Kepler [12] now offer automated execution of scientific workflows on remote computational resources in a more or less general way. However, these tools focus on expert users who know the behaviour of the computational codes used within the workflow, and are able to estimate the amount of computational resources needed by each task. The scheduling and mapping of the workflow on the computational resources are usually left to the cluster batch processing systems such as PBS[1] or Slurm[2].

These task schedulers provide their best effort to execute the tasks in the earliest possible time depending on the cluster workload and user/task priorities. However, what they cannot deal with is the execution parameters settings. If the user overestimates the amount of the computing resources, the tasks may be waiting in the queue for much longer time while making only little benefit from increased amount of resources, e.g., processor cores. On the other hand, underestimating these requirements may lead to the premature task termination due to exhausting the execution time.

This paper focuses on the heuristic-based selection of the execution parameters for a list of predefined computing codes used in the biomedical workflows supported by the k-Wave toolbox [18]. Since all binaries are fixed and known in advance, their performance characteristics such as strong and weak scaling can be automatically collected and used for prediction. Limiting the users in uploading their binaries also enables fine-grain performance tuning of the underlying codes for target machines and simplifies the workflows composition by the use of high-level processing blocks.

The next section describes the k-Plan system supporting the design of ultrasound workflows via a graphical user interface, and workflow offloading, scheduling, execution and monitoring using the k-Dispatch module. Section 3 describes a single pass optimization of the workflow execution parameters and related interpolation heuristics. Section 4 investigates the quality of interpolation for known and unknown tasks, and Sect. 5 concludes the paper.

---

[1] https://www.altair.com/pbs-works/.

[2] https://slurm.schedmd.com/.

2   Automatic Offloading of k-Wave Workflows

The k-Wave toolbox [18] is an open source Matlab toolbox designed for the time-domain simulation of acoustic waves propagating in tissues. The toolbox has a wide range of functionality, but at its heart is an advanced numerical model that can account for both linear and nonlinear wave propagation, an arbitrary distribution of heterogeneous material parameters, power law acoustic absorption and its thermal effects on the tissue. During recent years, k-Wave has attracted a lot of attention amongst biomedical physicists, ultrasonographers, neurologists and oncologists. Many k-Wave-based applications have been reported in photoacoustic breast screening [13], transcranial brain imaging [14], and high intensity focused ultrasound treatment planning for kidney [1,16], liver [7] or prostate tumour ablations [17].

However, all these applications require very intensive computations. During the last decade, the simulation core has been rewritten in C++ and parallelized by various technologies, such as OpenMP for shared memory systems [19], CUDA for GPU accelerated systems [10], and MPI for large distributed clusters [8]. These implementations now cover a wide range of ultrasound simulations in domains of various sizes reaching the limits of the top supercomputers.

To support clinicians in executing ultrasound workflows, a complex system called k-Plan [9], consisting of tree modules, is being developed, see Fig. 1:

1. TPM - Treatment Planning Module implements user front-end with the graphical user interface to compose the processing workflow. Advanced users may also use a Matlab interface or third-party applications.
2. DSM (k-Dispatch) - Dispatch Server Module is responsible for the workflow offloading to remote computing facilities. It also schedules particular tasks, estimates computing requirements, and monitors the workflow progress.
3. SEM - Simulation Execution Module covers the deployed binaries necessary to run particular tasks. Due to strict medical restrictions, all binaries have to be certified, thoroughly tested and properly deployed.

Although designed for the k-Wave toolbox, k-Dispatch remains as general as possible to support other applications and workflow types. User applications such as TPM communicate with k-Dispatch through the Web server, see Fig. 2. The Dispatch database maintains users and groups, their resource allocations, history of calculated and submitted workflows, available computing facilities, executable binaries with their performance characteristics, etc. Besides decoding the workflows, data transfers, monitoring and communication with remote computing facilities, the k-Dispatch core performs the optimization of the workflow execution parameters.

Users can create new ultrasound procedures by altering predefined workflow templates and packing them with the patient's data. Once delivered to k-Dispatch, the execution workflow is constructed from the provided input file. Next, the list of available computing resources is scanned to find a suitable one, e.g., the one with the lowest actual workload. Consequently, appropriate binaries for particular tasks are filled in to the workflow template according to the

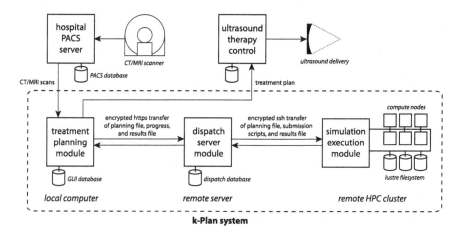

Fig. 1. Architecture of the k-Plan system. The dispatch server module (k-Dispatch) arranges for the workload scheduling, execution, monitoring and data transfers between client applications and computing facilities.

tasks input data size and available hardware. Since k-Dispatch knows the performance scaling of the given binaries, it can optimize the amount of computational resources (i.e., number of nodes) assigned to particular tasks and minimize several objectives such as cost, execution time and queuing time, see Algorithm 1.

After the tasks have been submitted to the computational queues, k-Dispatch keeps monitoring them, detects anomalies such as frozen/crashed jobs, and restarts them if necessary. After the workflow computation has been completed, the results are downloaded from the remote computing facility back to the k-Dispatch and the user is notified that the results are available for download.

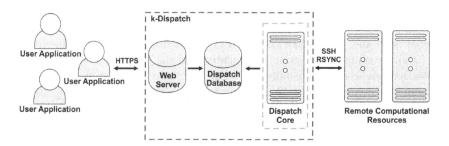

Fig. 2. k-Dispatch stands between user applications and remote computational resources. The communication with user applications is based on standard web services while the SSH protocol is used to communicate with remote computational resources. The dispatch core is responsible for the workflow submission, monitoring and other service mechanisms.

Algorithm 1: Adaptive execution planning algorithm

Presumptions :

1 Let $G = (V, E)$ be a workflow where V is a set of tasks and $E \subseteq V \times V$ is a set of task dependencies.

2 Let C be a set of active resource allocations with enough resources to satisfy the workflow G. It holds $C \subseteq A$, where A is a set of all allocations the user has got access to.

3 All executable binaries for supported task types available in a given allocation $a \in A$ are defined as $D \in (B_1, B_2, \ldots, B_N)$, where N is the number of task types within the workflow G, and $B_i = \{b_1, b_2, \ldots, b_M\}$ is the set of available binaries for a given task type. B_i may be an empty set.

4 Let $p : G \times C \times D \to \mathbb{R}^+$ be a price function returning the aggregated computational cost of the workflow G.

5 Let $t : G \times C \times D \to \mathbb{R}^+$ be a function returning the aggregated execution time of the workflow G. This value is calculated as a critical path through the workflow considering both the net execution time e and the queuing time q.

6 Let workflow evaluation f serving as quality metric be defined as $f = \alpha \cdot p + (1 - \alpha) \cdot t$, where α is a selectable ratio prioritizing the minimal computational cost or the execution time.

Algorithm   :

1 Create a workflow $G = (V, E)$ from the workflow template and input data.

2 Select a set of candidate allocations $C = \{c \in A^+ \mid c.status == active \land c.hours\_left > 0.0\}$.

3 Set appropriate execution parameters for all tasks and evaluate the workflow G for all combinations of candidate allocations C and binary executables D.

4 Return the best parameters for a given workflow G as $\operatorname{argmin}_{(c \in C, d \in D)} f(G)$.

---

3   Optimization of Workflow Execution Parameters

A typical course an ordinary user takes when executing a complex workflow is to use default execution parameters for each task, often consisting of one computational node and 24 h of wall time. If a task fails due to insufficient memory or time, another node or more time is allocated and the workflow restarted. Nevertheless, experienced users usually run a few benchmarks with various input sizes and number of nodes to create a strong scaling plot and predict the extent of computational resources for each task, which is the idea k-Dispatch has adopted.

In [9], three levels of workflow optimization were introduced. The naive one using the default execution parameters was implemented to compare k-Dispatch with other WMSs which use firmly set values directly provided by the users. This paper deals with a single-pass, task level optimization, processing each task independently. As we will show later, this is a viable solution with a linear time complexity providing sufficient results when execution cost and time is only considered. However, optimizing also for the queuing time requires a multi-pass, global optimization which may lead to an exponential time complexity, and needs a cluster simulator loaded with actual snapshots of cluster workload.

3.1   Single-Pass Optimization

The goal of the single-pass optimization is to independently find such execution parameters for each task i that minimize the workflow evaluation given by

$$f = \sum_{i=1}^{N} (\alpha * p_i + (1 - \alpha) * e_i) \tag{1}$$

where α is a weight preferring either execution cost or time, p is the execution cost and e is the net execution time. The queuing time is omitted here. Currently, the execution parameters to be optimized only cover the number of allocated nodes/cores and the execution time. Nevertheless, it is straightforward to extend the optimization to select the most suitable code, computational queue, node type (accelerated/fat/slim), etc.

Figure 3 illustrates the optimization of the task execution parameters as a black box with a task type and task input file provided by the workflow as the inputs. The task input file is parsed to extract information necessary to estimate the computational requirements. This information typically includes the size of the simulation domain, the simulation timespan, type of the medium, transducer definition, etc. Next, the collected performance data is searched to find similar records. Having a filtered out performance dataset, the plot of strong scaling can be constructed and several interpolation techniques can be used to estimate the task duration and cost for suitable amounts of resources. Once the best execution parameters are selected, the machine specific job scripts are generated and submitted to the computing queue. After the task has been properly finished, the performance data is used to update the performance database.

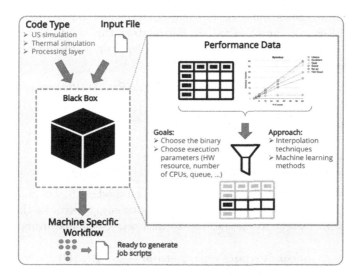

Fig. 3. Optimization of the execution parameter for a given task using a couple of heuristics and historically collected performance data for known code types.

3.2   Interpolation Heuristics

The goal of interpolation heuristics is to estimate the execution time and cost using the measured performance data from previous runs. Since the users are not limited in the size of the simulation domain and many other simulation parameters influencing the execution time, the performance data will never be complete.

There are three basic situations which may happen during the execution time and cost estimation:

1. The same simulation has been seen before. In such a case, the execution time and cost can be taken as a median value over multiple records stored in the database. If the values for particular amount of resources are unknown, an interpolation is used. Figure 4a shows this situation for four different domain sizes where the performance data are only known for 1, 2, 4 and 8 threads. The values for other numbers of threads have to be interpolated, see the question marks.
2. The simulation has not been seen before. In such a case, similar simulations are sought for in the database. First, the total number of grid points is calculated as a product of the dimension sizes. This may, however, unfavourably impact the estimation, since the actual shape does have an impact on the execution time, see Sect. 3.3. Next, all simulations with the number of grid points close to the one being estimated are selected. Finally, the execution time and cost are interpolated from the selected data. Figure 4b shows a situation where the performance data was only measured for 4 different domain sizes. The others have to be interpolated, see the yellow area.
3. The interpolation fails and it is necessary to use queue default wall time and amount of compute resources. This may happen if the simulation is too far from the known ones, or the interpolation method begins to oscillate and produces, e.g., negative values. Fortunately, this is a transient situation because as soon as the task is executed at least once, the measured values can be used next time.

Four interpolation methods offered by the SciPy [20] Python package were investigated in this paper:

- linear interpolation (LI),
- cubic spline interpolation (CS),
- nearest neighbour interpolation (NN),
- radial basis function interpolation (RBF).

As the quality measure for the interpolation methods, $L1$-, $L2$- and L-Infinity norms were used [6]. Additionally, the mean percentage error of the obtained data series with respect to the measured values was calculated using Eq. (2).

$$meanPercentError = mean(\frac{|a - b|}{|a|}) \times 100 \qquad (2)$$

where a is a vector of reference data series and b is a vector of interpolated data series.

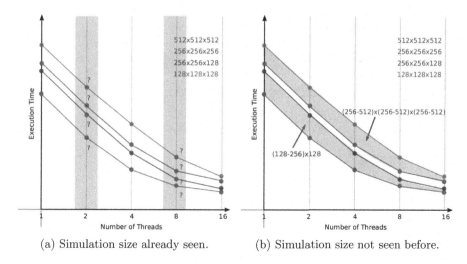

(a) Simulation size already seen.   (b) Simulation size not seen before.

Fig. 4. (a) The performance database misses data (highlighted in yellow) for some numbers of threads. The interpolation works with the corresponding strong scaling curves. (b) The performance database misses data for a range of domain sizes (highlighted by yellow areas). The interpolations works with several strong scaling curves from the close proximity. (Color figure online)

3.3   k-Wave Workflow Properties

A typical biomedical ultrasound workflow consists of several data processing and numerical simulation tasks. Together, they form a workflow with approximately 100 tasks. Figure 5 shows an example of the neurostimulation workflow. While the pre- and post-processor tasks require only a single computing node, the aberration correction, forward planning, and thermal simulations may employ various executables to run on a single node, a single GPU, or multiple nodes.

The simulation domain size and timespan is given by the subject anatomy, transducer position, and the ultrasound frequency. Considering small animal neurostimulations, the domain sizes can be as small as $162 \times 192 \times 128$ grid points with 3,000 simulation time steps. The move towards human patients may expand the simulation domain size up to $768 \times 900 \times 600$ grid points with 16,800 simulation time steps.

Figure 6 shows the performance behaviour of the distributed MPI version of the k-Wave toolbox for the largest practical domain normalised to a single simulation time step. The execution times were measured on the Anselm supercomputer using 1 to 16 compute nodes, each of which with 16 cores and 64 GB of memory. It can be seen that the performance scaling is not perfect with the maximum speed-up of 6.5 yielding the parallel efficiency of 40%. The yellow, green and orange dots mark the ideal amount of computational resources for three different values of the α parameters. If the execution time is preferred, the highest possible number of nodes is selected. On the other hand, if the execution

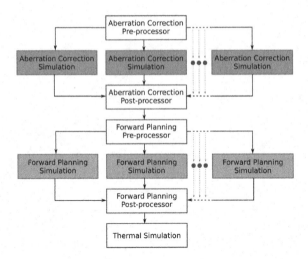

Fig. 5. A neurostimulation workflow consisting of several data processing and simulation tasks. The task dependencies are shown by the arrows, meaning the simulations depicted in red or blue may be executed concurrently. (Color figure online)

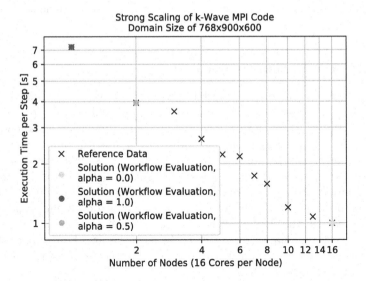

Fig. 6. Strong scaling of the MPI version of the k-Wave simulation in a domain consisting of $768 \times 900 \times 600$ grid points. The yellow, green and orange dots show the best number of nodes when minimizing the computational time, computational cost, or composite workflow evaluation, respectively. (Color figure online)

cost is preferred, a single node is selected. Finally, if both the time and cost have the same weight, two computing nodes looks as a good compromise.

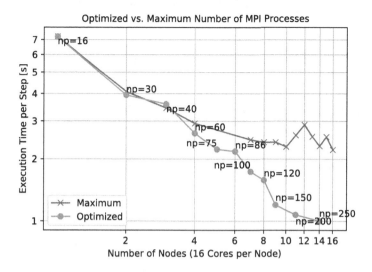

Fig. 7. Strong scaling of the MPI version of the k-Wave simulation on a domain of $768 \times 900 \times 600$ domain size executed with the maximum (blue line) and optimal (orange line) numbers of MPI processes (np). (Color figure online)

When working with the MPI version of the k-Wave toolbox, balanced work distribution must be paid attention to. Since the code uses a one-dimensional grid decomposition over the z dimension, and the grid is z-y transposed several times every time step, the z and y dimensions must be divisible by the number of MPI processes. Otherwise, the work is not balanced evenly and the code does not scale well. Figure 7 shows the scaling of the code executed with the maximum numbers of MPI processes for given number of nodes, and with reduced numbers of processes ensuring commensurability. It is obvious, the optimized numbers of processes yield higher performance.

3.4   Typical Problems of Performance Data Interpolations

The interpolation and extrapolation methods have several drawbacks that will be discussed in this section. We used measured performance data from Fig. 6 and tried to manually fit interpolation curves through the measured data.

Generally, the k-Wave codes have a linearithmic computation complexity $O(n \cdot log \cdot n)$ due to extensive use of 3D Fourier transform. However, the significant amount of communication stemming from the distributed FFT may lead to quadratic communication complexity. Moreover, the proper workload balancing as well as other restrictions imposed on the domain size make the scaling even more difficult to predict [5,8]. Therefore, there are significant differences in the course of the scaling curves at low and high numbers of threads/nodes.

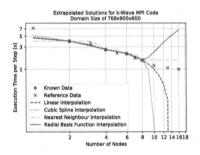

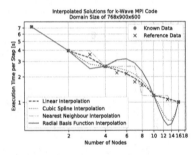

(a) Extrapolation based on the knowledge of scaling on 2, 4, 6 and 8 nodes.

(b) Interpolation based on the knowledge of scaling on 1, 2, 4, 10 and 16 nodes.

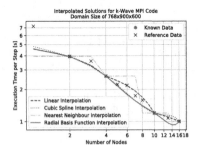

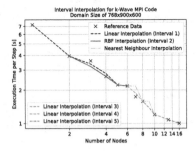

(c) Interpolation based on the knowledge of scaling on 2, 4, 10 and 16 nodes.

(d) Interval based interpolation, each interpolation uses 2 or 3 closest values.

Fig. 8. (a) Unsuccessful extrapolation trained on a small number of nodes. (b) and (c) Oscillation caused by distant known values. (d) Interval interpolation not suffering from the oscillations.

Figure 8a shows a poor attempt to extrapolation where the performance data is only known for 2, 4, 6 and 8 nodes. The estimation of the execution time for number of nodes above 8 is not acceptable. The linear extrapolation as well as cubic spline extrapolation predict much shorter execution time. Nevertheless, the code scales much worse for higher number of nodes because the communication component starts to dominate. The nearest neighbour extrapolation could be used as the worst case, however, Fig. 6 suggests that the performance can even deteriorate with higher number of compute nodes. Finally, the radial basis interpolation does not produce meaningful predictions.

Figure 8b and c point out the need to abide appropriate interval between known values to eliminate oscillations. Figure 8b uses an additional value for one node compared to Fig. 8c. This value is usually an outlier causing unintended oscillations since having no communication. To reduce them, several interpolations may be performed on smaller intervals. The impact of this technique is shown in Fig. 8d, where the scaling data is divided into 5 intervals of 2 to 3 values. However, it is not clear how to determine the interval size automatically.

4   Experimental Results

This section describes performed experiments and the results. The experiments show the application of the selected interpolation methods in order to autonomously find the suitable execution parameters.

Due to the necessity of collecting an extensive performance dataset, we limited ourselves to only consider the OpenMP k-Wave implementation running on a single node, however, with various numbers of threads. The execution cost was then calculated as a product of the execution time and the number of processor cores used. In principle, similar results are expected to be obtained for the CUDA implementation of k-Wave. On the other hand, the MPI version poses more restrictions and may feature different results, see Sect. 3.4.

The performance data collected for the OpenMP code was obtained on Anselm with 16 cores per node, and Salomon with 24 cores per node. The performance data was divided into the training and testing datasets both of which containing over 6,500 records of the aberration correction k-Wave simulation running over 24 different domain sizes (32^3 to 512^3 grid points) and with various number of threads.

4.1   Comparison of Interpolation Techniques for Known Simulation

We first investigated the behaviour of all four interpolation techniques on the known domain size of 512^3 grid points. The first experiment used 6 known execution times from the Anselm cluster measured for 1, 2, 4, 8, 12 and 16 threads. Table 1 and Fig. 9 show the course of the interpolation functions. It can be seen that the linear and cubic spline interpolation methods reached less than 3% mean error. The linear interpolation can be thought of as a pessimistic one since overestimating the execution times. Although this may lead to a bit longer queuing times, it is safer than underestimation produced by the cubic spline interpolation, which may lead to premature termination of the simulation. The nearest neighbour interpolation shows significantly worse accuracy as well as the radial basis function interpolation deeply oscillating, especially for high numbers of threads.

The second experiment extended the number of measured values and also included the Salomon cluster. For Anselm, the performance data was extracted from the database for 1, 2, 4, 5, 8, 10, 13, and 15 threads, while for Salomon

Table 1. Comparison of selected interpolation methods for domain size of 512^3 grid points domain size and 6 known values measured on Anselm.

Interpolation method	L1-Norm	L2-Norm	L2-Infinity Norm	Mean error [%]
Linear	1.27	0.59	0.46	2.89
Cubic Spline	0.93	0.42	0.35	2.29
Nearest Neighbour	4.60	2.40	2.06	9.85
Radial Basis Function	3.41	1.37	0.79	8.77

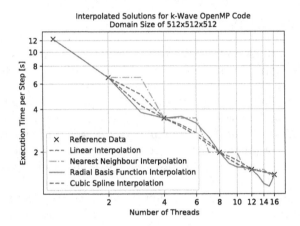

Fig. 9. Comparison of various interpolation techniques for the OpenMP implementation of k-Wave running on Anselm with a domain size of 512^3 grid points.

the set was further extended by performance data for 17, 20, 22, and 24 threads. This covers 50% of all possible thread numbers usable on both clusters. The domain size remained the same (512^3 grid points).

Tables 2 and 3 show significant improvement in the prediction accuracy. The mean error produced by the linear interpolation was reduced from 2.89% to 1.81%, and 1.27% on Anselm and Salomon, respectively. Even better results were achieved for the cubic spline interpolation which produced estimation with only 1.23% and 1.12% error. Even the other interpolation methods improved

Table 2. Comparison of selected interpolation methods for domain size of 512^3 grid points domain size and 8 known values measured on Anselm.

Interpolation method	L1-Norm	L2-Norm	L2-Infinity Norm	Mean error [%]
Linear	0.80	0.45	0.41	1.81
Cubic Spline	0.56	0.38	0.37	1.23
Nearest Neighbour	2.99	2.03	1.95	5.70
Radial Basis Function	1.61	0.90	0.67	4.67

Table 3. Comparison of selected interpolation methods for domain size of 512^3 grid points domain size and 12 known values measured on Salomon.

Interpolation method	L1-Norm	L2-Norm	L2-Infinity Norm	Mean error [%]
Linear	0.62	0.33	0.29	1.27
Cubic Spline	0.60	0.40	0.39	1.12
Nearest Neighbour	2.73	1.71	1.63	4.68
Radial Basis Function	1.08	0.69	0.66	2.01

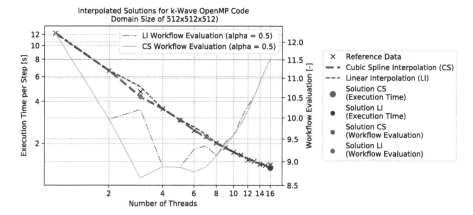

Fig. 10. Estimation of the best execution configuration according to the workflow evaluation function for a domain size of 512^3 grid points on the Anselm cluster produced by linear and cubic spline interpolation.

the error close to or below 5%. This can be considered as a very good result since there is always a slight variation in execution times between different runs caused by the underlying cluster workload (mainly network and I/O parts), and variations in clock frequency amongst different cluster nodes.

Figure 10 illustrates the result of the interpolation for linear and cubic spline interpolation for the extended training set, and the domain size of 512^3 grid points. The curves show a very good agreement without any significant oscillations. The orange and grey curves are the visualizations of the workflow evaluation functions with $\alpha = 0.5$. If looking for the fastest solution, both the linear and cubic spline interpolations predict 16 threads to be the best solution. In the case the combined workflow evaluation metric is minimized, 3 and 5 threads are predicted as best compromises by the cubic spline and linear interpolations, respectively.

4.2 Comparison of Interpolation Techniques for Unknown Simulations

This set of experiments evaluates the capabilities of the proposed interpolation methods to estimate the execution time for simulations that have not been seen before. In this case, the closest simulations in terms of the total number of grid points are used to fit the interpolation curves. Since the results were similar for both clusters, we only present measurements on Anselm.

Three different unknown domain sizes were tested:

1. Tested simulation size 256×224^2, training set containing simulations of 224^3, $256^2 \times 224$, and $224^2 \times 192$ grid points.
2. Tested simulation size $160^2 \times 128$, training set containing simulations of 144^3, 160^3, and 132×128^2 grid points.

3. Tested simulation size 144^3, training set containing simulations of 160^3, 160×128^2, and 132×128^2 grid points.

Figure 11 shows the results of selected interpolations on the first two simulation domains. Both linear and cubic spline interpolations show a very close agreement with the reference data stored in the testing set. As Tables 4 and 5 quantify, the mean error for the biggest domain reaches 4.7% and 3.1% for linear and cubic spline interpolations, respectively. For the smaller domain, the error decreases to 1.75% and 2.25%. Interestingly, the cubic spline produces slightly

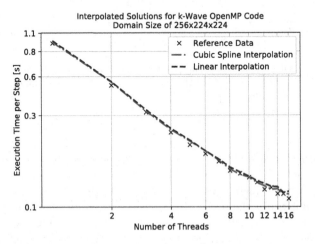

(a) Domain size of 256×224^2 grid points

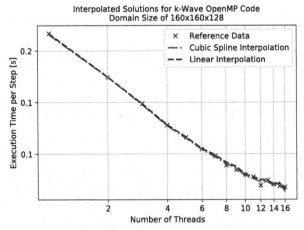

(b) Domain size of $160^2 \times 128$ grid points

Fig. 11. Comparisons of linear and cubic spline interpolation methods for unknown domain sizes. The reference data points are used for the error evaluation.

Table 4. Comparisons of selected interpolation methods for an unknown domain sizes of 256×224^2 grid points.

Interpolation method	L1-Norm	L2-Norm	L2-Infinity Norm	Mean error [%]
Linear	0.17	0.056	0.034	4.724
Cubic Spline	0.11	0.037	0.025	3.073
Nearest Neighbour	0.84	0.271	0.191	22.35
Radial Basis Function	352	99.26	47.99	11492

Table 5. Comparisons of selected interpolation methods for an unknown domain sizes of $160^2 \times 128$ grid points.

Interpolation method	L1-Norm	L2-Norm	L2-Infinity Norm	Mean error [%]
Linear	0.015	0.005	0.003	1.75
Cubic Spline	0.023	0.007	0.005	2.25
Nearest Neighbour	0.252	0.089	0.068	17.7
Radial Basis Function	0.371	0.121	0.089	29.2

worse estimations here. The nearest neighbour interpolation gives much worse estimation with a mean error of 22% and 18% for those two cases. Finally, the radial basis interpolation appears to be unusable for the largest domain. The extreme error is caused by high oscillations. In case of the medium-sized domain, the error decreases to 29%. Unfortunately, this still exceeds acceptable values.

The smallest domain size of interest suffers from very poor results which are summarized in Table 6 and Fig. 12. The only usable estimations are provided by the linear interpolation, however, with a mean error of 16%. The cubic spline completely fails in this case while the best estimation is surprising provided by the nearest neighbour interpolation. The radial basis interpolation also fails on this domain size. The overestimation is very likely caused by a small domain size when a single grid can fit into L3 cache memory leading to much faster execution of the Fourier transforms and overall algorithm speed-up. On the other hand, even overestimation by 200% may be thought of as acceptable considering such a simulation is executed within 2 min using 16 threads.

Table 6. Comparisons of linear and cubic spline interpolation methods for an unknown domain sizes of 144^3 grid points.

Interpolation method	L1-Norm	L2-Norm	L2-Infinity Norm	Mean difference [%]
Linear	0.196	0.061	0.041	15.99
Cubic Spline	2.080	0.527	0.185	212.6
Nearest Neighbour	0.177	0.064	0.050	13.40
Radial Basis Function	4.050	1.024	0.356	416.8

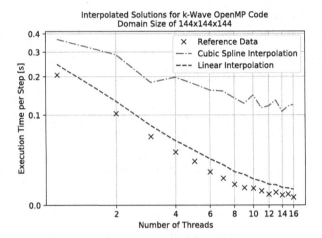

Fig. 12. Comparisons of linear and cubic spline interpolation methods for unknown domain size of 144^3. The reference data points are used for the error evaluation.

5   Conclusions

The need for offloading complex scientific workflows to cluster and cloud environment is ubiquitous. k-Dispatch is a workflow management system providing automated execution, planning and monitoring of biomedical workflows composed of k-Wave ultrasound and thermal simulations. Its interface enables connection of various user applications and unifies the access to different computational resources.

One of the key challenges in automated execution of complex workflows is the proper setting of execution parameters for particular tasks. Since the end users have no or very limited knowledge about the amount of computational resources to be allocated for each task, it is necessary to provide as good estimation as possible based on the performance characteristics of particular codes and actual input data. Unsuitable values may lead to long queueing times or early tasks termination due to exhausted time allocation.

This paper has presented a single pass algorithm traversing the workflow and optimizing the execution parameters for every task independently. For every task, the input file is inspected, the task parameters retrieved, and the performance database searched for similar ones. If there is a direct match, the execution time and cost are loaded for known execution parameters, i.e., number of compute nodes, GPUs, processor cores, etc. Missing values may be filled in using interpolation techniques. However, if the task parameters have not been seen before, the interpolation is used to estimate the execution time and cost using a training set composed of tasks with similar parameters.

Four different interpolation techniques have been investigated. When the task parameters have been seen before, the cubic spline interpolation showed the best results with mean error between 1.12% and 2.29%. In the case the task parameters have not been seen before, the linear interpolation showed the best results. Depending on the similarity of the records found in the performance database, the mean error varies between 1.17% and 15%. It should be noted that the highest error showed up only for very small tasks where the overestimation of execution time or cost do not play a significant role.

5.1   Future Work

Future work will be focused on multi-pass optimization of workflow execution parameters. The goal is to minimize not only the execution time and cost but also the queuing times. This however requires the knowledge of the actual cluster workload and queues occupancy as well as a cluster simulator to quickly estimate the queuing times for the whole workflow under different execution parameters. We are considering the adaptation of the ALEA simulator [11] to match the scheduling algorithms and hardware configurations of IT4Innovations clusters, and the characteristics of the k-Wave workflows.

We would also like to implement more sophisticated heuristics to select an appropriate number of compute nodes as well as optimal number of MPI processes for large simulations to avoid performance penalizations. Consequently, we would like to study machine learning methods since we expect to have collected large performance dataset, and perform experiments on both, artificial and real-world workflows.

Acknowledgement. This work was supported by the FIT-S-17-3994 Advanced parallel and embedded computer systems project. This work was supported by The Ministry of Education, Youth and Sports from the National Programme of Sustainability (NPU II) project IT4Innovations excellence in science - LQ1602 and by the IT4Innovations infrastructure which is supported from the Large Infrastructures for Research, Experimental Development and Innovations project IT4Innovations National Supercomputing Center - LM2015070. This work was supported by the Engineering and Physical Sciences Research Council, United Kingdom, grant numbers EP/L020262/1, EP/M011119/1, EP/P008860/1, and EP/S026371/1.

References

1. Abbas, A., Coussios, C., Cleveland, R.: Patient specific simulation of HIFU kidney tumour ablation, vol. 2018, pp. 5709–5712 (July 2018). https://doi.org/10.1109/EMBC.2018.8513647
2. Su, M.H.: Pegasus: a framework for mapping complex scientific workflows onto distributed systems. Sci. Program. **13**, 219–237 (2005)
3. Deelman, E., et al.: Pegasus: a workflow management system for science automation. Fut. Gener. Comput. Syst. **46**, 17–35 (2014)
4. Foster, I.: Globus toolkit version 4: software for service-oriented systems. J. Comput. Sci. Technol **21**(4), 513–520 (2006). https://doi.org/10.1007/s11390-006-0513-y
5. Frigo, M., Johnson, S.: The design and implementation of FFTW3. Proc. IEEE **93**(2), 216–231 (2005). https://doi.org/10.1109/JPROC.2004.840301

6. Gradshtein, I.S.: Table of Integrals, Series, and Products. Academic Press, San Diego (2000)
7. Grisey, A., Yon, S., Letort, V., Lafitte, P.: Simulation of high-intensity focused ultrasound lesions in presence of boiling. J. Ther. Ultrasound (2016). https://doi.org/10.1186/S40349-016-0056-9
8. Jaros, J., Rendell, A.P., Treeby, B.E.: Full-wave nonlinear ultrasound simulation on distributed clusters with applications in high-intensity focused ultrasound. Int. J. High Perform. Comput. Appl. **30**(2), 137–155 (2016). https://doi.org/10.1177/1094342015581024
9. Jaros, M., Treeby, B.E., Georgiou, P., Jaros, J.: k-Dispatch: a workflow management system for the automated execution of biomedical ultrasound simulations on remote computing resources. In: Proceedings of the Platform for Advanced Scientific Computing Conference, PASC 2020. Association for Computing Machinery, New York (2020). https://doi.org/10.1145/3394277.3401854
10. Kadlubiak, K., Jaros, J., Treeby, B.E.: GPU-accelerated simulation of elastic wave propagation. In: 2018 International Conference on High Performance Computing & Simulation (HPCS), pp. 188–195. IEEE (July 2018). https://doi.org/10.1109/HPCS.2018.00044
11. Klusacek, D., Toth, S., Podolnikova, G.: Complex job scheduling simulations with Alea 4. CEUR Workshop Proc. **1828**, 53–59 (2017). https://doi.org/10.1145/1235
12. Ludäscher, B., et al.: Scientific workflow management and the Kepler system. Concurrency Comput. Pract. Exp. **18**(10), 1039–1065 (2006). https://doi.org/10.1002/cpe.994
13. Manohar, S., Dantuma, M.: Current and future trends in photoacoustic breast imaging. Photoacoustics **16**, 100134 (2019). https://doi.org/10.1016/j.pacs.2019.04.004
14. Mohammadi, L., Behnam, H., Tavakkoli, J., Avanaki, M.R.: Skull's photoacoustic attenuation and dispersion modeling with deterministic ray-tracing: towards real-time aberration correction. Sensors (Switzerland) (2019). https://doi.org/10.3390/s19020345
15. Robert, Y.: Task graph scheduling. In: Padua, D. (ed.) Encyclopedia of Parallel Computing. Springer, Boston (2011). https://doi.org/10.1007/978-0-387-09766-4_42
16. Suomi, V., Jaros, J., Treeby, B., Cleveland, R.: Nonlinear 3-D simulation of high-intensity focused ultrasound therapy in the Kidney. Conf. Proc. IEEE Eng. Med. Biol. Soc., 5648–5651 (2016). IEEE. https://doi.org/10.1109/EMBC.2016.7592008
17. Suomi, V., et al.: Transurethral ultrasound therapy of the prostate in the presence of calcifications: a simulation study. Med. Phys. **45**, 4793–4805 (2018). https://doi.org/10.1002/mp.13183
18. Treeby, B.E., Cox, B.T.: k-Wave: MATLAB toolbox for the simulation and reconstruction of photoacoustic wave-fields. J. Biomed. Opt. **15**(2), 021314 (2010)
19. Treeby, B.E., Jaros, J., Rendell, A.P., Cox, B.T.: Modeling nonlinear ultrasound propagation in heterogeneous media with power law absorption using a k-space pseudospectral method. J. Acoust. Soc. Am. **131**(6), 4324–4336 (2012). https://doi.org/10.1121/1.4712021
20. Virtanen, P., Gommers, R., Oliphant, T.E., et al.: SciPy 1.0: fundamental algorithms for scientific computing in Python. Nat. Meth. **17**, 261–272 (2020)

Analysis and Visualization of the Dynamic Behavior of HPC Applications

Ondrej Vysocky[(✉)] [ID], Ivo Peterek, Martin Beseda[ID], Matej Spetko[ID],
David Ulcak, and Lubomir Riha[ID]

IT4Innovations National Supercomputing Center,
VSB - Technical University of Ostrava, Ostrava, Czech Republic
{ondrej.vysocky,ivo.peterek,martin.beseda,matej.spetko,david.ulcak,
lubomir.riha}@vsb.cz

Abstract. The behavior of a parallel application can be presented in many ways, but performance visualization tools usually focus on communication graphs and runtime of processes or threads in specific (groups of) functions. A different approach is required when searching for the optimal configuration of tunable parameters, for which it is necessary to run the application several times and compare the resource consumption of these runs. We present RADAR visualizer, a tool that was originally developed to analyze such measurements and to detect the optimal configuration for each instrumented part of the code. In this case, the optimum was defined as the minimum energy consumption of the whole application, but any other metric can be defined.

RADAR visualizer presents the application behavior in several graphical representations and tables including the amount of savings that can be reached. Together with our MERIC library, we provide a complete toolchain for HPC application behavior monitoring, data analysis, and graphical representation. The final part is performing dynamic tuning (applying optimal settings for each region during the application runtime) for the production runs of the analyzed application.

Keywords: MERIC · READEX · Energy efficient computing · Performance analysis · HPC

1 Introduction

On the way to exascale supercomputing, the high performance computing community faces the problem of the power and energy consumption of these systems. Consequently, many projects are focusing on solutions to optimize performance under a specified power budget or maximize energy savings with minimal impact on application performance [3,4,16]. Amongst these was a Horizon 2020 project called READEX [14,17]. The main idea of the project is based on splitting an application into parts, for which we expect different hardware requirements.

© Springer Nature Switzerland AG 2021
T. Kozubek et al. (Eds.): HPCSE 2019, LNCS 12456, pp. 135–149, 2021.
https://doi.org/10.1007/978-3-030-67077-1_8

The READEX tools tune available knobs to fit the application needs for each of its instrumented parts to avoid wasting resources. We speak about dynamic tuning, as opposed to static tuning, where only one configuration is set at the beginning of the application execution.

As a member of the READEX project consortium, we have implemented the lightweight C++ library MERIC [9,21] for parallel application resource consumption measurement and hardware parameter tuning during application runtime according to the READEX approach. This means that MERIC stores an application profile under various hardware settings into an output record. This record is searched for settings that consumed a minimal amount of energy, runtime, and potentially other metrics. During an application run, MERIC may tune CPU core (CF) and uncore (UnCF) frequency (on Intel processors it is frequency of subsystems in the physical processor package that is shared by multiple processor cores e.g., L3 cache or on-chip ring interconnect), RAPL power cap limit, and number of active OpenMP threads if available on the target architecture.

MERIC output can be analyzed using the python tool RADAR [10,21] with a graphical user interface RADAR visualizer. RADAR provides various representations to compare the application resource consumption under different parameter configurations and identifies which configuration of each region will bring the maximum energy savings. There are many tools for application behavior analysis, usually with their own profile visualization graphical tool, such as Score-P [11] and Scalasca [6] that use Cube, TAU [20] and its ParaProf, and Extrae [19] with a Paraver tool. These tools were originally implemented to provide insight into an application behavior within a single run to compare resource consumption between parallel processes or threads. This is not true for the Open|SpeedShop tool, which allows comparison of up to eight database files, where each contains information about one run of the analyzed application [18], which is useful when a new improved implementation is done.

Our use case requires comparison of tens of runs of an application, each with a different configuration. Moreover, each part of the application may have a unique configuration during a single run. We evaluate these runs and automatically identify the optimal configuration of each region of the application.

In contrast to the previously mentioned HPC application analysis tools, we are not focused on presenting any details of application behavior below the socket level, since energy measurement systems provide the consumption for each socket (e.g. Intel RAPL counters [8], HDEEM [7]) and in some cases for the whole node only (e.g. DiG [12]).

This paper follows up our previous work [10,21] presenting RADAR as a command-line Python tool, which generates a LaTeX document with series of tables and graphs. The RADAR tool has been replaced with the RADAR visualizer GUI application, that brings not only much higher user-friendliness, but new visualisation elements and data analysis algorithms too.

2   Understanding Application Behavior

The RADAR visualizer provides a graphical representation of an analyzed application behavior. It both visualizes the measured data, and also does its own analysis to find the optimal configuration of the application's regions. The results of the analysis are mostly presented in tables, explained into more detail later. The optimal configuration can be also stored into a configuration file for MERIC, to tune the application during its production runs.

The READEX approach, as well as all other energy-saving approaches, reduces available resources, which obviously may cause an overhead. The reduction in performance stems from the balancing of energy savings against the subsequent performance penalty. In the case of the READEX approach it is possible for the production phase to exclude all configurations that cause any overhead, or specify an acceptable limit.

To show an example of the dynamic behavior of a parallel application we will present selected parts of the Lattice Boltzmann method (LBM) application analysis. This computational fluid dynamics simulation application is a good example since it is a simple application that exhibits much exploitable dynamism [1, 22]. The main loop contains two regions that behave very differently. First, the Propagate part moves populations between lattice-sites and is strongly memory-bound, then the compute intensive Collide region follows, which recomputes the values of populations at each site. For the test, we selected a 4096 × 4096 points lattices size and 25 iterations of the main loop. Such a high amount of iterations is not required for the analysis, but can help us identify any inter-dynamism within the regions. In this paper we use the application to depict the features of our tools only; the achievable energy savings of this application have been already presented in [2].

The presented values were measured on a single node of the TU Dresden Taurus supercomputer powered by two Intel Xeon CPUs E5-2680v3 (codename Haswell) with 12 cores each, which operates on a core frequency of 1.2–3.30 GHz (2.5 GHz nominal frequency, in the following Sects. 2.6 GHz will represent the turbo frequency) and 1.2–3.0 GHz uncore frequency. Taurus nodes accommodate the HDEEM system, which was used for energy measurement. An exhaustive space state search was applied with a 200 MHz step size for both CPU core and uncore frequencies, and even numbers from 12 up to 24 active threads were used.

The RADAR visualizer does not have a single main window. When running the tool a panel of buttons is displayed, each opening a new window with one of the following representations of the application behavior.

2.1   Charts, Plots and Heatmaps

Regions' Call-Path Structure. First of all, we should start with a look at the whole application as MERIC sees it during the analysis. The application has been divided into regions by MERIC's instrumentation, which has been inserted manually into the source code of the analyzed application or automatically directly to the application binary. When running the instrumented application MERIC measures and records resources consumption for each region.

The RADAR visualizer provides a chart that shows the structure of calls of the regions in the application. This RADAR window allows us to store the chart as an image, or into yEd graph format[1], which is useful for visualization of complex graphs. The advantage of this chart format is that the RADAR visualizer does not have to specify the exact location of each node, but the yEd Graph editor provides several advanced automatic layout transformations.

Based on the call-path structure, or for example because of the region's too short runtime, the user may remove some of the regions from the analysis (the region remains in the code) to make the process faster and the graphical presentations of the results more lucid. The interface provides the option of automatic region selection using time a filter or selection of all nested regions of one top-level region.

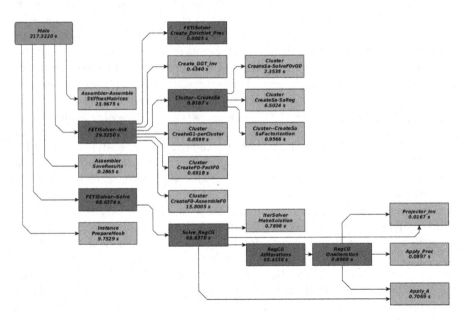

Fig. 1. ESPRESO FETI solver regions' callpath graph in a yEd graph. (Color figure online)

The LBM application has a very simple structure with an initialization region and one loop which contains Propagate and Collide regions. Therefore, in this case, we use an example of a complex graph generated from the RADAR visualizer in yEd format for the ESPRESO FETI solver [15]. The chart is shown in Fig. 1. Under the Main region, only the last-level regions are selected for analysis (yellow color), from which one region is deselected (red color) for its short runtime.

---

[1] yEd Graph editor can be download from https://www.yworks.com/yed.

Direct Values. Once we have selected the regions to analyze, we may evaluate the behavior of the whole application and also each of its regions for various configurations.

Even though the list of tuned parameters might be in general unlimited, we provide visualization of the data in two-dimensional visual elements only to keep the level of lucidity, and to be able to produce a printed report. The user must specify which parameters should be shown on the x or y axes. For the rest of the parameters, we select the optimal values of these parameters and inform the user of the values.

The RADAR visualizer provides two visual elements, plot and heatmap, both showing the same values measured by MERIC under specific configurations. However, from the plot it is easier to see trends, and in the case of the heatmap it is easier to present exact values. Furthermore, by clicking on a specific cell of a heatmap, the user can get information about the values, from which the value in the cell has been calculated. This might be useful for detecting outliers in a measurement.

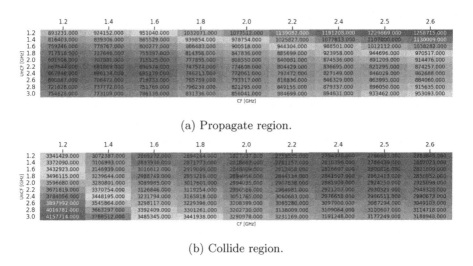

(a) Propagate region.

(b) Collide region.

Fig. 2. Heatmap representation of energy consumption in millijoules of the regions Propagate and Collide when using 12 OpenMP threads, and CPU core and uncore frequencies are tuned.

Figure 2 shows such heatmaps for energy consumption of the Propagate and Collide regions when 12 threads are used. From these heatmaps, it is observable that minimal energy consumption is reached via completely different configurations, presenting the highest energy consumption in red cells at the opposite sides of the heatmaps. To maximally reduce the energy consumption of the LBM application it is necessary to switch from high CPU uncore and low core frequencies, which are optimal for the Propagate region, to low uncore frequency and high core frequency when the Collide region starts, and then back to the previous configuration. For example, we can read from the heatmaps the difference in minimum and maximum energy consumption; over 590 J for the Propagate region and 1 370 J for the Collide region.

There is almost zero impact on Propagate region runtime when the CPU core frequency is underclocked (Fig. 3). In this case, only CPU uncore frequency influences the region runtime, which again confirms that the region is highly memory-bound.

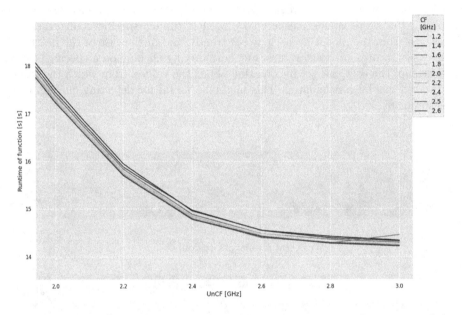

Fig. 3. Impact of the CPU core and uncore frequency tuning on the runtime of the Propagate region. Zoomed to the 2.0–3.0 GHz uncore frequency.

Power Samples Visualization. Some energy measurement systems (those supported by MERIC include HDEEM and DiG) provide not only energy consumption in a specified time period but also access to power samples that have been taken by the system. On user request the MERIC library may also store these power samples in the output file[2]. These power samples hold much important data that is worth investigating, especially in the default settings of the

---

[2] By default, MERIC does not store the power samples because it creates a much larger output. Both HDEEM and DiG work normally on a 1 kHz sampling frequency for the blade, which means a thousand entries per second of the measurement. HDEEM also has sensors on specific parts of the node (e.g. each CPU or each memory channel) and measures them on 100 Hz sampling frequency, so in cases where these samples are also included, the output size rises accordingly.

CPU, when the default governor tunes the frequencies itself, otherwise some changes in the power consumption might be hidden.

The power timeline of both CPUs, and two of the four most significant memory channels of the LBM application single run when no specific hardware settings are applied (the system scaled the frequencies automatically) is presented in Fig. 4. In this chart the initialization and the following 25 iterations of the loop with Propagate and Collide regions are clearly visible.

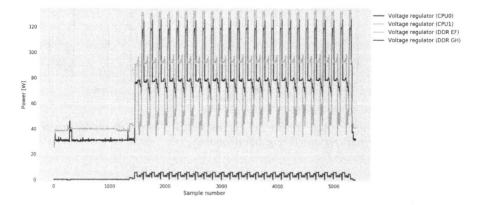

Fig. 4. Power timeline of a single run of the LBM application.

From our observation, the power consumption may slightly jitter around some value, even when the node is under the same workload. Together with a high frequency of sampling, the plot might be less informative about the trends in power consumption. To overcome this problem we provide a possibility to smooth the graph by applying a filter that inserts smoothed values calculated from the surrounding n-samples into the graph. The smoothing is based on the Savitzky–Golay filter without differentiating the data [13] where the user may specify the window size (number of surrounding samples, default 11) and the order of the polynomial (default 2) used to fit the power samples.

As an example of a smoothed power timeline, we show a fragment of power consumption of the whole computational node during five iterations of the application node in Fig. 5. The regions' calls are inserted separately, so we can clearly identify calls of the Propagate (blue) and Collide (red) regions.

When visualizing regions with more than one call using line plot, we add an extra 0 W sample at the beginning and end of each region call (except at the beginning of the first and end of the last call). In Fig. 5 you can see extra lines going down and up, otherwise, there would be unwanted lines connecting the first power sample with the last power sample of the previous region call.

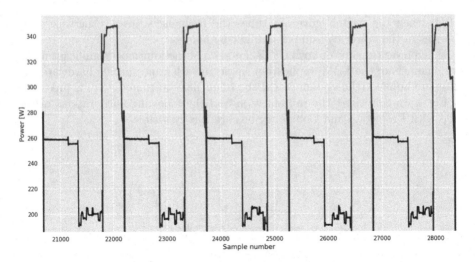

Fig. 5. Smoothed node power timeline of the main LBM loop containing Propagate (blue) and Collide (red) regions (Color figure online).

The power timeline of a region carries important information; it may show that the region should be split into two or more smaller regions if there are continuous clusters of power samples with approximately the same value. If such a region is not covered with one or more nested regions, we lose an opportunity to exploit the dynamism. The RADAR visualizer provides a cluster analysis of the power samples and informs the user that a region that is not covered by nested ones should be split into smaller regions to obtain the maximum possible savings.

The cluster analysis performed on the LBM run suggests splitting the region Propagate into two regions of approximately the same size. The output can be seen in Fig. 6, where the analysis was performed both on data from the blade power sensor and one of the memory channels power sensors, respectively. The figure is magnified to include four calls of the Propagate region only, where two different parts of the region can be seen for each of the region's calls in both sub-figures. Standalone dots (samples) represent outliers not included in any of the clusters.

The cluster analysis itself utilizes the DBSCAN algorithm [5], employing a weighted Euclidean distance (1) as a metric. By setting $(w_1, w_2) = (0.001, 0.85)$ we managed to re-scale the x-axis to values of the same order as the y-axis and slightly contract the y-axis. Thus we managed to diminish the oscillations while preserving a significant difference in the power samples among regions. Furthermore, a standard deviation s (2) multiplied by 0.85 was employed as a guess for a sufficient neighborhood radius ϵ. The algorithm is set to identify all regions of the minimum 100 ms size.

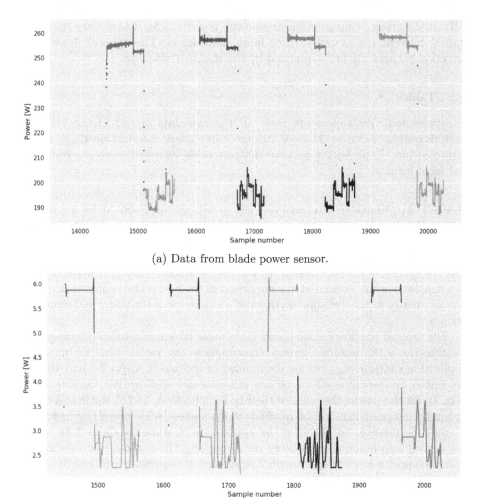

(a) Data from blade power sensor.

(b) Data from DDR EF power sensor.

Fig. 6. Detail at four consecutive calls of the Propagate region. The cluster analysis performed by DBSCAN splits each call into two separate parts.

$$d_{wE}(a, b) = \sum_{i=1}^{2} w_i(a_i - b_i)^2, \quad a, b \in \mathbb{R}^2 \tag{1}$$

$$s = \sqrt{\frac{1}{n-1} \sum_{i=1}^{n} (x_i - \overline{x})}, \quad n \in \mathbb{N} \tag{2}$$

Another kind of dynamism can be detected when the cluster analysis does not compare power samples during a single region call but within all the region's calls. If the energy consumption or runtime is not stable, it may vary together with a different region call-path or input parameters. This type of dynamism detection is currently not supported by the RADAR visualizer.

2.2 Tables

The presented visual elements show all the raw data stored in the MERIC analysis output, however, there is still more important information that can be extracted too. To present it we have chosen a table representation as a compact and lucid form.

Overall Application Summary. First of all, the user should be interested in the *Overall application summary* table, which provides summary information about the application resource consumption in its default settings, if the analysis applied a static tuning then also in the best static configuration, and also in the best configuration for each region if dynamic parameters tuning is applied. The dynamic savings are evaluated in comparison to the best static configuration; the overall savings of the dynamic savings are a summary of the static and dynamic savings.

The Overall application summary table takes into account that you may not want to reach the minimal energy consumption but rather tune for minimal application runtime (e.g. reduce the number of threads in a specific part that is influenced by the numa effect) or any other measured objective. Consequently (Fig. 7), for the production runs of the LBM application, 23.7 % static and 6.2 % dynamic energy savings can be reached, which makes 28.5 % overall savings, and it could be possible to optimize the application runtime to about 6.8 s without taking into account the energy consumption. The bottom row informs us that the application will run about 2.76 s longer if the energy optimal dynamic configuration were applied.

	Default settings	Default values	Best static configuration	Static savings	Dynamic savings
Energy consumption [J], Blade summary - hdeem	24thrds, 3.0GHz UnCF, 2.6GHz CF	12227.70J	12thrds, 2.0GHz UnCF, 1.8GHz CF	2905.79J (23.76%)	578.26J of 9321.91J (6.20%)
Runtime of function [s], Blade summary - hdeem	24thrds, 3.0GHz UnCF, 2.6GHz CF	53.42s	16thrds, 3.0GHz UnCF, 2.6GHz CF	1.79s (3.36%)	4.97s of 51.63s (9.62%)
Run-time change with the energy optimal settings	-2.76s (96.85 % of default time)				

Fig. 7. Comparison of the time and energy consumption of the LBM application run in the default, the optimal static, and the best dynamic configuration.

Average Program Start. The window *Average program start* lists all the regions nested in the main region and presents their impact on the application. For each region, the user has information about the percentage region size from the consumption of the selected resource (time, energy, etc.) point of view. Percents express the coverage of the region in the main region. This column follows the optimal static configuration of the main region (it is the same for all the nested regions) and resource consumption of each region in this configuration. The best dynamic configuration is obviously an individual for each region. Also, the subsequent column named *Value* presents the region's consumption in the best dynamic configuration. Finally, dynamic savings gives us information on how much we save in comparison to the best static configuration of the main region.

Region	% of 1 phase	Best static configuration	Value	Best dynamic configuration	Value	Dynamic savings
Collide	38.65	12 thrds, 2.0 GHz UnCF, 1.8 GHz CF	3382.67 J	18 thrds, 1.8 GHz UnCF, 1.4 GHz CF	2906.22 J	476.45 J (14.08%)
Init	15.55	12 thrds, 2.0 GHz UnCF, 1.8 GHz CF	1360.59 J	18 thrds, 1.2 GHz UnCF, 1.8 GHz CF	1293.07 J	67.52 J (4.96%)
Propagate	45.80	12 thrds, 2.0 GHz UnCF, 1.8 GHz CF	4008.90 J	12 thrds, 2.2 GHz UnCF, 1.2 GHz CF	3980.52 J	28.38 J (0.71%)
Total value for static tuning for significant regions	3382.67 + 1360.59 + 4008.90 = 8752.16 J					
Total savings for dynamic tuning for significant regions	476.45 + 67.52 + 28.38 = 572.35 J of 8752.16 J (6.54%)					
Dynamic savings for application runtime	572.35 J of 9321.91 J (6.14%)					
Total value after savings	8749.56 J (71.56% of 12227.70 J)					
Run-time change with the energy optimal settings against the default time settings (region-wise):	+0.85s,(108.47%); +0.85s,(107.62%); -5.39s,(80.81%);					

Fig. 8. Table of the selected nested regions of the LBM application and their optimal dynamic configurations to reach minimal energy consumption.

When having a list of the nested regions we may see which regions consume the most resources and focus our interest on them. In Fig. 8 we may see minimal energy savings gained from dynamic tuning of the Propagate region. The region takes over 45 % of the application runtime, so its optimal configuration almost matches the optimal static configuration. On the other hand, the Collide region achieves better energy consumption when 18 threads are active with 1.8 GHz uncore and 1.4 GHz core CPU frequency. Such configuration brings 14.1 % of savings for the Collide region and another 5 % of the energy consumed by the Init region can be saved if it is switched to its optimal configuration.

Both absolute and percentage savings that come from the dynamic tuning of these regions are summarized in the last five rows of the table. The table presents energy savings only, but the user may also select other objectives to be presented in another table.

Nested Region Table. The last type of available table focuses again on just one selected region behavior only. This table is designed for regions located in an instrumented loop (such instrumentation is called a phase region in the READEX terminology) and compares how the behavior of such regions changes during the iterations.

Phase ID	1	2	3
Default Energy consumption [J]	160.78	154.00	160.11
% per 1 phase	54.39	53.50	54.34
Per phase optimal settings	12 threads, 2.0 GHz UnCF, 1.6 GHz CF	12 threads, 2.0 GHz UnCF, 2.4 GHz CF	12 threads, 2.2 GHz UnCF, 2.2 GHz CF
Dynamic savings [J]	5.58	3.31	5.24
Dynamic savings [%]	3.47	2.15	3.27
Def. and eng. optima diff[s]	-0.27	-0.23	-0.27

Fig. 9. Table comparing the first three calls of the Propagate region.

Figure 9 shows only the beginning of the whole Nested region table. Due to the size of the table, we show only the Energy consumption during the first three iterations (Phase ID) of the loop. In the table, we can see how the optimal configuration changes during iterations, and what the resource consumption is.

The goal of this kind of region behavior representation is the detection of regions that behave differently every N iterations (e.g. storing output to a file every ten iterations), which means that we should identify different optimal configurations accordingly. In the future version of the RADAR visualizer, we will also prepare a plot representation and automatic detection of this kind of dynamism.

3   Conclusion

In this paper we present the RADAR visualizer, which provides graphical representation and analysis of parallel applications' resource consumption measured by the MERIC library, and tunes the application according the READEX approach based on exploiting application dynamism. The application must be measured in several configurations. These runs are then compared by our tool, which identifies the optimal configuration for each instrumented part of the code. The measurements are presented using several available visual representations; a graph of regions' callpath, a plot and heatmap comparing resource consumption in various configurations, a power consumption timeline, and tables presenting the available savings.

These visual elements can be directly exported into several image formats, and can be arranged and combined to produce a final LATEX based report for the presentation of the application behavior to the user or application developer.

Future work includes extending the RADAR visualizer with new graphical representations (e.g. n-dimensional plots to visualize data from multiple computational nodes side by side) and new algorithms for data analysis (e.g. add a restriction for maximum impact of the configuration on the application runtime) that can help us reach the maximum savings from various kinds of application dynamism.

Acknowledgment. This work was supported by The Ministry of Education, Youth and Sports from the Large Infrastructures for Research, Experimental Development and Innovations project IT4Innovations National Supercomputing Center LM2015070.

This work was supported by The Ministry of Education, Youth and Sports from the Large Infrastructures for Research, Experimental Development and Innovations project „e-Infrastructure CZ - LM2018140".

This work was supported by the Moravian-Silesian Region from the programme "Support of science and research in the Moravian-Silesian Region 2017" (RRC/10/2017).

This work was partially supported by the SGC grant No. SP2019/59 "Infrastructure research and development of HPC libraries and tools", VŠB - Technical University of Ostrava, Czech Republic.

References

1. Calore, E., Gabbana, A., Schifano, S.F., Tripiccione, R.: Evaluation of DVFS techniques on modern HPC processors and accelerators for energy aware applications. Concurr. Comput. Pract. Exp. **29**(12), e4143 (2017). https://doi.org/10.1002/cpe.4143

2. Calore, E., Gabbana, A., Schifano, S.F., Tripiccione, R.: Energy-efficiency tuning of a lattice Boltzmann simulation using MERIC. In: Wyrzykowski, R., Deelman, E., Dongarra, J., Karczewski, K. (eds.) PPAM 2019. LNCS, vol. 12044, pp. 169–180. Springer, Cham (2020). https://doi.org/10.1007/978-3-030-43222-5_15

3. Cesarini, D., Bartolini, A., Bonfà, P., Cavazzoni, C., Benini, L.: COUNTDOWN - three, two, one, low power! A run-time library for energy saving in MPI communication primitives (2018). CoRR abs/1806.07258. http://arxiv.org/abs/1806.07258

4. Eastep, J., et al.: Global extensible open power manager: a vehicle for HPC community collaboration on co-designed energy management solutions. In: Kunkel, J.M., Yokota, R., Balaji, P., Keyes, D. (eds.) ISC 2017. LNCS, vol. 10266, pp. 394–412. Springer, Cham (2017). https://doi.org/10.1007/978-3-319-58667-0_21

5. Ester, M., Kriegel, H.P., Sander, J., Xu, X.: A density-based algorithm for discovering clusters in large spatial databases with noise. In: Proceedings of the Second International Conference on Knowledge Discovery and Data Mining, KDD1996, pp. 226–231. AAAI Press (1996)

6. Geimer, M., Wolf, F., Wylie, B.J.N., Ábrahám, E., Becker, D., Mohr, B.: The SCALASCA performance toolset architecture. Concurr. Comput. Pract. Exper. **22**(6), 702–719 (2010). https://doi.org/10.1002/cpe.v22:6
7. Hackenberg, D., Ilsche, T., Schuchart, J., Schöne, R., Nagel, W.E., Simon, M., Georgiou, Y.: HDEEM: High definition energy efficiency monitoring. In: 2014 Energy Efficient Supercomputing Workshop, pp. 1–10 (2014).https://doi.org/10.1109/E2SC.2014.13
8. Hähnel, M., Döbel, B., Völp, M., Härtig, H.: Measuring energy consumption for short code paths using RAPL. SIGMETRICS Perform. Eval. Rev. **40**(3), 13–17 (2012). https://doi.org/10.1145/2425248.2425252
9. IT4Innovations: MERIC library. https://code.it4i.cz/vys0053/meric. Accessed 21 Apr 2019
10. IT4Innovations: READEX RADAR library. https://code.it4i.cz/bes0030/readex-radar. Accessed 21 Aug 2019
11. Knüpfer, A., et al.: Score-p: a joint performance measurement run-time infrastructure for periscope, SCALASCA, TAU, and VAMPIR. In: Brunst, H., Müller, M.S., Nagel, W.E., Resch, M.M. (eds.) Tools for High Performance Computing 2011, pp. 79–91. Springer, Berlin Heidelberg (2012)
12. Libri, A., Bartolini, A., Benini, L.: Dwarf in a giant: Enabling scalable, high-resolution HPC energy monitoring for real-time profiling and analytics (2018). CoRR abs/1806.02698: http://arxiv.org/abs/1806.02698
13. Persson, P.O., Strang, G.: Smoothing by Savitzky-Golay and Legendre filters. In: Rosenthal, J., Gilliam, D.S. (eds.) Mathematical Systems Theory in Biology, Communications, Computation, and Finance, pp. 301–315. Springer, New York, New York, NY (2003)
14. READEX: Horizon 2020 READEX project (2018). https://www.readex.eu
15. Riha, L., et al.: A massively parallel and memory-efficient fem toolbox with a hybrid total FETI solver with accelerator support. Int. J. High Perform. Comput. Appl. 0(0), 1094342018798452 (0). https://doi.org/10.1177/1094342018798452
16. Rountree, B., Lowenthal, D.K., de Supinski, B.R., Schulz, M., Freeh, V.W., Bletsch, T.K.: Adagio: making DVS practical for complex HPC applications. In: Proceedings of the 23rd International Conference on Supercomputing, ICS 2009, pp. 460–469. ACM, New York, NY, USA (2009). https://doi.org/10.1145/1542275.1542340
17. Schuchart, J., et al.: The READEX formalism for automatic tuning for energy efficiency. Computing **99**(8), 727–745 (2017). https://doi.org/10.1007/s00607-016-0532-7
18. Schulz, M., Galarowicz, J., Maghrak, D., Hachfeld, W., Montoya, D., Cranford, S.: Open Speedshop: an open source infrastructure for parallel performance analysis. Sci. Program. **16**(2–3), 105–121 (2008). https://doi.org/10.1155/2008/713705
19. Servat, H., Llort, G., Huck, K., Giménez, J., Labarta, J.: Framework for a productive performance optimization. Parallel Comput. **39**(8), 336–353 (2013). https://doi.org/10.1016/j.parco.2013.05.004
20. Shende, S.S., Malony, A.D.: The TAU parallel performance system. Int. J. High Perform. Comput. Appl. **20**(2), 287–311 (2006). https://doi.org/10.1177/1094342006064482

21. Vysocky, O., et al.: Evaluation of the HPC applications dynamic behavior in terms of energy consumption. In: Proceedings of the Fifth International Conference on Parallel, Distributed, Grid and Cloud Computing for Engineering, pp. 1–19 (2017), paper 3, 2017. https://doi.org/10.4203/ccp.111.3
22. Vysocky, O., Riha, L., Zapletal, J.: A simple framework for energy efficiency evaluation and hardware parameter tuning with modular support for different HPC platforms. In: Proceedings of the Eighth International Conference on Advanced Communications and Computation, pp. 25–30. IARIA (2018). http://www.thinkmind.org/index.php?view=article&articleid=infocomp_2018_2_10_68005

A Convenient Graph Connectedness for Digital Imagery

Josef Šlapal$^{(\boxtimes)}$

IT4Innovations Centre of Excellence, Brno University of Technology,
Brno, Czech Republic
slapal@fme.vutbr.cz

Abstract. In a simple undirected graph, we introduce a special connectedness induced by a set of paths of length 2. We focus on the 8-adjacency graph (with the vertex set $\mathbb{Z}^2$) and study the connectedness induced by a certain set of paths of length 2 in the graph. For this connectedness, we prove a digital Jordan curve theorem by determining the Jordan curves, i.e., the circles in the graph that separate $\mathbb{Z}^2$ into exactly two connected components. These Jordan curves are shown to have an advantage over those given by the Khalimsky topology on $\mathbb{Z}^2$.

Keywords: Simple undirected graph · Connectedness · Digital plane · Khalimsky topology · Jordan curve theorem.

1 Introduction

In our increasingly digital world, digital images become an integral part of our everyday life. They play an extremely important role in scientific data visualization and this is the main reason for studying them in this paper.

In digital geometry for two-dimensional (2D for short) computer imagery, we usually replace pixels of a computer screen by their center points so that the screen is then represented by a finite section of the digital plane $\mathbb{Z}^2$. But, instead of such a section, we work with the whole digital plane $\mathbb{Z}^2$. A 2D black and white digital image is then a finite subset of $\mathbb{Z}^2$ and its elements are called black points. The remaining elements of $\mathbb{Z}^2$, called white points, form the background of the image. One of the basic problems of 2D digital image analysis and processing is to find a convenient connectedness structure for the digital plane $\mathbb{Z}^2$. Since digital images are simply digital approximations of the real ones, a connectedness being convenient means that the digital plane provided with such a structure behaves in much the same way as the Euclidean plane. In particular, it is required that such a structure allows for a digital analogue of the Jordan curve theorem (recall that the classical Jordan curve theorem states that a Jordan, i.e., simple closed, curve in the Euclidean plane separates this plane into exactly two connected components). In digital images, digital Jordan curves represent borders of objects imaged and, therefore, play an important role in solving numerous problems such as pattern recognition, memory usage compression, image reconstruction, etc.

© Springer Nature Switzerland AG 2021
T. Kozubek et al. (Eds.): HPCSE 2019, LNCS 12456, pp. 150–162, 2021.
https://doi.org/10.1007/978-3-030-67077-1_9

The classical, graph-theoretic approach to solving the problem of providing the digital plane with a convenient connectedness structure is based on using the well-known 4- and 8-adjacency graphs (see e.g. [7,8,12–14,17]). A disadvantage of this approach is that neither of the two graphs itself allows for a digital Jordan curve theorem so that a combination of them has to be used. Therefore, in [3], a new, topological approach to the problem was proposed based on employing a single structure, the so-called called Khalimsky topology, to obtain a convenient connectedness in the digital plane $\mathbb{Z}^2$. The topological approach was then developed by many authors - see, e.g., [4–6,9–11,15,16].

The Khalimsky topology has the property that its connectedness coincides with the connectedness in a simple undirected graph with the vertex set $\mathbb{Z}^2$, namely the connectedness graph of the topology. Thus, to equip the digital plane with a convenient connectedness structure, this graph, rather than the Khalimsky topology itself, may be used. A drawback of this approach is that Jordan curves in the (connectedness graph of the) Khalimsky topology may never turn at the acute angle $\frac{\pi}{4}$. It would, therefore, be useful to find some new, more convenient structures on $\mathbb{Z}^2$ that would allow Jordan curves to turn, at some points, at the acute angle $\frac{\pi}{4}$. In the present note, to obtain such a convenient structure, we employ the 8-adjacency graph with connectedness given by a certain set of paths of length 2 in the graph. For this connectedness, we prove a digital Jordan curve theorem to show that the graph with the set of paths provides a convenient structure on the digital plane for the study and processing of digital images.

2   Preliminaries

For the graph-theoretic concepts used see, for instance, [1]. By a *graph* we always mean an undirected simple graph without loops, hence and ordred pair (V, E) of sets where $E \subseteq \{\{a, b\};\ a, b \in V,\ a \neq b\}$. The elements of V are called *vertices* and those of E are called edges of the graph. If $\{a, b\} \in E$, then the vertices a and b are said to be *adjacent* and the edge $\{a, b\}$ is said to *join* the vertices a and b. For an arbitrary vertex $a \in V$, we denote by $E(a)$ the set of all vertices adjacent to a, i.e., $E(a) = \{b \in V;\ \{a, b\} \in E\}$. Clearly, $\{a, b\} \in E$ if and only if $b \in E(a)$ or, equivalently, $a \in E(b)$. Thus, the set E of edges of a graph may be given by determining the set $E(a)$ for every $a \in V$.

As usual, we graphically represent graphs by thinking of vertices as points and edges as line segments whose end points are just the vertices they join.

A graph (U, F) is called a *subgraph* of a graph (V, E) if $U \subseteq V$ and $F \subseteq E$. If, moreover, $F = E \cap \{\{a, b\};\ a, b \in U\}$, then (U, F) is said to be an *induced subgraph* of (V, E) being denoted briefly by U. A subgraph (U, F) of (V, E) is called a *factor* of (V, E) if $U = V$.

Recall that a *walk* in a graph (V, E) is a finite sequence $(a_i|\ i \leq n) = (a_0, a_1, ..., a_n)$, n a non-negative integer, of vertices such that $\{a_{i-1}, a_i\} \in E$ whenever $i \in \{1, 2, ...n\}$. If all vertices a_i, $i \in \{0, 1, ..., n\}$, are pairwise different, then the walk $(a_i|\ i \leq n)$ is said to be a *path* and the number n is called the *length* of the path. Thus, also a single vertex is considered to be a path (of length 0). A *circle* in (V, E) is any walk $(a_i|\ i \le n)$ with $n > 2$ such that $(a_i|\ i < n)$ is a path and $a_0 = a_n$. A subset $X \subseteq V$ is said to be *connected* if, for every pair $a, b \in X$, there is a path $(a_i|\ i \le n)$ such that $a_0 = a$, $a_n = b$ and $a_i \in X$ for all $i \in \{0, 1, ..., n\}$. A maximal (with respect to set inclusion) connected subset of V is called a *component* of the graph (V, E).

A nonempty, finite and connected subset C of V is said to be a *simple closed curve* in (V, E) if the set $E(a) \cap C$ has two elements for every $a \in C$. Clearly, every simple closed curve is a circle. A simple closed curve in (V, E) is called a *Jordan curve* in (V, E) if it separates the set V into exactly two components, i.e., if the induced subgraph $V - C$ of (V, E) has exactly two components.

For every point $(x, y) \in \mathbb{Z}^2$, we put $A_4(x, y) = \{(x + i, y + j);\ i, j \in \{-1, 0, 1\},\ ij = 0,\ i + j \ne 0\}$ and $A_8(x, y) = A_4(x, y) \cup \{(x + i, y + j);\ i, j \in \{-1, 1\}\}$. The points of $A_4(x, y)$ and $A_8(x, y)$ are said to be 4-adjacent and 8-adjacent to (x, y), respectively. The graphs $(\mathbb{Z}^2, A_4)$ and $(\mathbb{Z}^2, A_8)$ are called the *4-adjacency graph* and *8-adjacency graph*, respectively, and are demonstrated in Fig. 1.

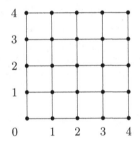 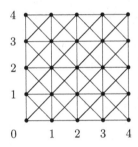

Fig. 1. Portions of the 4- and 8-adjacency graphs.

In digital image processing, the 4-adjacency and 8-adjacency graphs are the most frequently used structures on the digital plane. But, since the late 1980's, another structure on $\mathbb{Z}^2$ has been used too, namely the Khalimsky topology [3]. It is the product of two copies of the topology on $\mathbb{Z}$ given by the subbase $\{\{2k - 1, 2k, 2k + 1\};\ k \in \mathbb{Z}\}$ (for the basic concepts of general topology see [2]). Recall that, given a topology $\mathcal{T}$ on a set X, the *connectedness graph* of $\mathcal{T}$ is the graph with the vertex set X such that a pair of different points $x, y \in X$ is adjacent if and only if $\{x, y\}$ is a connected subset of the space $(X, \mathcal{T})$. Since the Khalimsky topology is an Alexandroff topology (which means that the closure operator in the topology is completely additive), the connectedness in the Khalimsky topological space coincide with the connectedness in the connectedness graph of the Khalimsky topology. We will call the connectedness graph of the Khalimsky topology briefly the *Khalimsky graph*. The Khalimsky graph is the graph $(\mathbb{Z}^2, K)$ such that, for any $(x, y) \in \mathbb{Z}^2$,

$$K(x,y) = \begin{cases} A_8(x,y) & \text{if } x \text{ and } y \text{ have the same parity,} \\ A_4(x,y) & \text{if } x \text{ and } y \text{ have different parities.} \end{cases}$$

A portion of the Khalimsky graph is demonstrated in Fig. 2. It is obvious that the graph is a factor of the 8-adjacency graph.

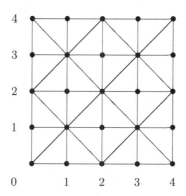

Fig. 2. A portion of the Khalimsky graph.

The famous Jordan curve theorem proved for the Khalimsky topology in [3] may be formulated as follows:

Theorem 1. *In the Khalimsky graph, every simple closed curve with at least four points is a Jordan curve.*

We denote by $(\mathbb{Z}^2, L)$ the factor of the Khalimsky graph $(\mathbb{Z}^2, K)$ given by $L = K - \bigcup\{\{(x,y),(z,t)\}; \ (x,y) \in \mathbb{Z}^2, \ x \text{ and } y \text{ are odd and } (z,t) \in A_4(x,y)\}$. A portion of the graph $(\mathbb{Z}^2, L)$ is demonstrated in Fig. 3.

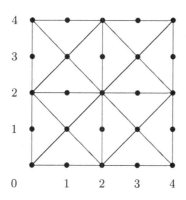

Fig. 3. A portion of the graph $(\mathbb{Z}^2, L)$.

The below corollary immediately follows from Theorem 1:

Corollary 1. *Every circle in the graph* $(\mathbb{Z}^2, L)$ *which does not turn, at any of its points, at the acute angle* $\frac{\pi}{4}$ *is a Jordan curve in the connectedness graph of the Khalimsky topology.*

It is readily verified that a simple closed curve (and thus also a Jordan curve) in the Khalimsky graph may never turn at the acute angle $\frac{\pi}{4}$. It could therefore be useful to replace the Khalimsky topology (Khalimsky graph) with some more convenient structure on $\mathbb{Z}^2$, another factor of the 8-adjacency graph, that would allow Jordan curves to turn at the acute angle $\frac{\pi}{4}$ at some points. And this is what we will do in the next section.

3   8-Adjacency Graph with a Set of Paths of Length 2

In the 8-adjacency graph $(\mathbb{Z}^2, A_8)$, the set $A_8(x, y)$ provides the digital plane $\mathbb{Z}^2$ with a natural concept of neighborhood of any point $(x, y) \in \mathbb{Z}^2$. Therefore, it would be desirable to use the graph for structuring the digital plane. But the usual concept of connectedness in the 8-adjacency graph does not alow for a digital Jordan curve theorem. To solve this problem, we employ another concept of connectedness.

Definition 1. Let (V, E) be a graph, $\mathcal{B}$ a set of paths of length 2 in the graph, and n a nonnegative integer. A sequence $C = (c_i \mid i \leq n)$ of elements of V is called a $\mathcal{B}$-*walk* if one of the the following three conditions is satisfied for every $i \in \{0, 1, ...n - 1\}$:

(i) There exists $(a_0, a_1, a_2) \in \mathcal{B}$ such that $\{c_i, c_{i+1}\} = \{a_0, a_1\}$,
(ii) $i > 0$ and there exists $(a_0, a_1, a_2) \in \mathcal{B}$ such that $c_{i-1} = a_0$, $c_i = a_1$, and $c_{i+1} = a_2$,
(iii) $i < n - 1$ and there exists $(a_0, a_1, a_2) \in \mathcal{B}$ such that $c_i = a_2$, $c_{i+1} = a_1$, and $c_{i+2} = a_0$.

A $\mathcal{B}$-walk $(c_i \mid i \leq n)$ with the property that $n \geq 2$ and $c_i = c_j \Leftrightarrow \{i, j\} = \{0, n\}$ is said to be a $\mathcal{B}$-*circle*.

Observe that, if $(x_0, x_1, ..., x_n)$ is a $\mathcal{B}$-walk, then $(x_n, x_{n-1}, ..., x_0)$ is a $\mathcal{B}$-walk, too (so that $\mathcal{B}$-walks are closed under reversion) and, if $(x_i \mid i \leq m)$ and $(y_i \mid i \leq p)$ are $\mathcal{B}$-walks with $x_m = y_0$, then, putting $z_i = x_i$ for all $i \leq m$ and $z_i = y_{i-m}$ for all i with $m \leq i \leq m + p$, we get a $\mathcal{B}$-walk $(z_i \mid i \leq m + p)$ (so that $\mathcal{B}$-walks are closed under composition).

Given a set $\mathcal{B}$ of paths of length 2 in a graph (V, E), a subset $A \subseteq V$ is said to be $\mathcal{B}$-*connected* if, for every pair $a, b \in A$, there is a $\mathcal{B}$-walk $(c_i \mid i \leq n)$ such that $c_0 = a$, $c_n = b$ and $c_i \in A$ for all $i \in \{0, 1, ..., n\}$. A maximal (with respect to set inclusion) $\mathcal{B}$-connected subset of V is called a $\mathcal{B}$-*component* of (V, E).

Definition 2. Let $\mathcal{B}$ be a set of paths of length 2 in a graph (V, E). A nonempty, finite and $\mathcal{B}$-connected subset J of V is said to be a $\mathcal{B}$-*simple closed curve* if every element $(a_0, a_1, a_2) \in \mathcal{B}$ with $\{a_0, a_1\} \subseteq J$ satisfies $a_2 \in J$ and every $c \in J$ fulfills one of the following two conditions:

(1) There are exactly two elements $(a_0, a_1, a_2) \in \mathcal{B}$ satisfying both $\{a_0, a_1, a_2\} \subseteq J$ and $c \in \{a_0, a_2\}$ and there is no element $(b_0, b_1, b_2) \in \mathcal{B}$ satisfying both $\{b_0, b_1, b_2\} \subseteq J$ and $c = b_1$.

(2) There is exactly one element $(b_0, b_1, b_2) \in \mathcal{B}$ satisfying both $\{b_0, b_1, b_2\} \subseteq J$ and $c = b_1$ and there is no element $(a_0, a_1, a_2) \in \mathcal{B}$ satisfying both $\{a_0, a_1, a_2\} \subseteq J$ and $c \in \{a_0, a_2\}$.

Clearly, every $\mathcal{B}$-simple closed curve is a $\mathcal{B}$-circle.

Definition 3. Let $\mathcal{B}$ be a set of paths of length 2 in a graph (V, E). A $\mathcal{B}$-simple closed curve J is called a $\mathcal{B}$-*Jordan curve* if the subset $V - J \subseteq V$ consists (i.e., is the union) of exactly two $\mathcal{B}$-components.

From now on, $\mathcal{B}$ will denote the set of paths of length 2 in the 8-adjacency graph given as follows: For every $((x_i, y_i)| \ i \leq 2)$ such that $(x_i, y_i) \in \mathbb{Z}^2$ for every $i \leq 2$, $((x_i, y_i)| \ i \leq 2) \in \mathcal{B}$ if and only if one of the following eight conditions is satisfied:

(1) $x_0 = x_1 = x_2$ and there is $k \in \mathbb{Z}$ such that $y_i = 4k + i$ for all $i \leq 2$,
(2) $x_0 = x_1 = x_2$ and there is $k \in \mathbb{Z}$ such that $y_i = 4k - i$ for all $i \leq 2$,
(3) $y_0 = y_1 = y_2$ and there is $k \in \mathbb{Z}$ such that $x_i = 4k + i$ for all $i \leq 2$,
(4) $y_0 = y_1 = y_2$ and there is $k \in \mathbb{Z}$ such that $x_i = 4k - i$ for all $i \leq 2$,
(5) there is $k \in \mathbb{Z}$ such that $x_i = 4k + i$ for all $i \leq 2$ and there is $l \in \mathbb{Z}$ such that $y_i = 4l + i$ for all $i \leq 2$,
(6) there is $k \in \mathbb{Z}$ such that $x_i = 4k + i$ for all $i \leq 2$ and there is $l \in \mathbb{Z}$ such that $y_i = 4l - i$ for all $i \leq 2$,
(7) there is $k \in \mathbb{Z}$ such that $x_i = 4k - i$ for all $i \leq 2$ and there is $l \in \mathbb{Z}$ such that $y_i = 4l + i$ for all $i \leq 2$,
(8) there is $k \in \mathbb{Z}$ such that $x_i = 4k - i$ for all $i \leq 2$ and there is $l \in \mathbb{Z}$ such that $y_i = 4l - i$ for all $i \leq 2$.

A portion of $\mathcal{B}$ is shown in Fig. 4. The paths of length 2, i.e., the ordered triples, belonging to $\mathcal{B}$ are represented by line segments oriented from first to last terms.

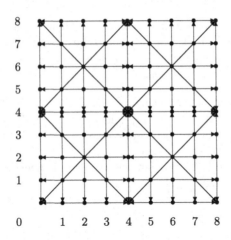

Fig. 4. A portion of the set $\mathcal{B}$.

Further, we denote by $(\mathbb{Z}^2, A)$ the factor of the 8-adjacency graph given as follows:

$$
A(x,y) = \begin{cases}
A_8(x,y) \text{ if } (x,y) = (4k, 4l), \ k, l \in \mathbb{Z}, \\
A_8(x,y) - A_4(x,y) \text{ if } (x,y) = (4k+2, 4l+2), \ k, l \in \mathbb{Z}, \\
\{(x-1,y), (x+1,y)\} \text{ if } (x,y) = (4k+i, 4l), \ k, l \in \mathbb{Z}, \\
\qquad\qquad\qquad\qquad i \in \{1,2,3\}, \\
\{(x,y-1), (x,y+1)\} \text{ if } (x,y) = (4k, 4l+i), \ k, l \in \mathbb{Z}, \\
\qquad\qquad\qquad\qquad i \in \{1,2,3\}, \\
\{(x-1,y-1), (x+1,y+1)\} \text{ if } (x,y) = (4k+i, 4l+i), \\
\qquad\qquad\qquad\qquad k, l \in \mathbb{Z}, \ i \in \{-1,1\}, \\
\{(x-1,y+1), (x+1,y-1)\} \text{ if } (x,y) = (4k+i, 4l-i), \\
\qquad\qquad\qquad\qquad k, l \in \mathbb{Z}, \ i \in \{-1,1\}, \\
\emptyset \text{ otherwise.}
\end{cases}
$$

A portion of the graph $(\mathbb{Z}^2, A)$ is demonstrated by Fig. 5.

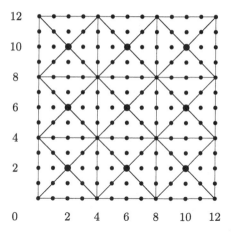

Fig. 5. A portion of the graph $(\mathbb{Z}^2, A)$.

Theorem 2. *Every circle in the graph $(\mathbb{Z}^2, A)$ that does not turn at any point $(4k + 2, 4l + 2)$, $k, l \in \mathbb{Z}$ (i.e, any point denoted by a bold dot in Fig. 5) is a $\mathcal{B}$-Jordan curve.*

Proof. Clearly, every circle in the graph $(\mathbb{Z}^2, A)$ is a $\mathcal{B}$-simple closed curve. Let $z = (x, y) \in \mathbb{Z}^2$ be a point such that $x = 4k + p$ and $y = 4l + q$ for some $k, l, p, q \in \mathbb{Z}$ with $pq = \pm 1$. Then, we define the *fundamental triangle* $T(z)$ to be the fifteen-point subset of $\mathbb{Z}^2$ given as follows:

$$
T(z) = \begin{cases}
\{(r, s) \in \mathbb{Z}^2; 4k \leq r \leq 4k + 4, \ 4l \leq s \leq 4l + 4k + 4 - r\} \\
\quad \text{if } x = 4k + 1 \text{ and } y = 4l + 1 \text{ for some } k, l \in \mathbb{Z}, \\
\{(r, s) \in \mathbb{Z}^2; \ 4k \leq r \leq 4k + 4, \ 4l \leq s \leq 4l + r - 4k\} \\
\quad \text{if } x = 4k + 3 \text{ and } y = 4l + 1 \text{ for some } k, l \in \mathbb{Z}, \\
\{(r, s) \in \mathbb{Z}^2; \ 4k \leq r \leq 4l + 4, \ 4l + 4k + 4 - r \leq s \leq 4l + 4\} \\
\quad \text{if } x = 4k + 3 \text{ and } y = 4l + 3 \text{ for some } k, l \in \mathbb{Z}, \\
\{(r, s) \in \mathbb{Z}^2; \ 4k \leq r \leq 4k + 4, \ 4l + r - 4k \leq s \leq 4l + 4\} \\
\quad \text{if } x = 4k + 1 \text{ and } y = 4l + 3 \text{ for some } k, l \in \mathbb{Z}.
\end{cases}
$$

Graphically, every fundamental triangle $T(z)$ consists of fifteen points and forms a right triangle obtained from a 4×4-square by dividing it by a diagonal. More precisely, each of the two diagonals divides the square into just two fundamental triangles having a common hypotenuse coinciding with the diagonal. In every fundamental triangle $T(z)$, the point z is one of the three internal points of the triangle. The (four types of) fundamental triangles are demonstrated by the below figure:

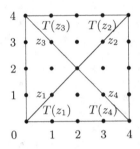

Given a fundamental triangle, we speak about its sides - it is clear from the above picture which sets are understood to be the sides (note that each side consists of five points and that two different fundamental triangles may have at most one side in common).

Now, one can easily see that

(1) every fundamental triangle is $\mathcal{B}$-connected and so is every subset of $\mathbb{Z}^2$ obtained by subtracting, from a fundamental triangle, some of its sides.

Consequently,

(2) if S_1, S_2 are fundamental triangles having a common side D, then the set $(S_1 \cup S_2) - M$ is $\mathcal{B}$-connected whenever M is the union of some sides of S_1 or S_2 different from D.

It is also evident that,

(3) whenever S_1, S_2 are different fundamental triangles with a common side D and $X \subseteq S_1 \cup S_2$ is a $\mathcal{B}$-connected subset with $X \cap S_1 \neq \emptyset \neq X \cap S_2$, we have $X \cap D \neq \emptyset$.

We will show that, for every circle C in the graph $(\mathbb{Z}^2, A)$ which does not turn at any point $(4k + 2, 4l + 2)$, $k, l \in \mathbb{Z}$, there are sequences $\mathcal{S}_F, \mathcal{S}_I$ of fundamental triangles, $\mathcal{S}_F$ finite and $\mathcal{S}_I$ infinite, such that, whenever $\mathcal{S} \in \{\mathcal{S}_F, \mathcal{S}_I\}$, the following two conditions are satisfied:

(a) Each member of $\mathcal{S}$, excluding the first one, has a common side with at least one of its predecessors.
(b) C is the union of those sides of fundamental triangles in $\mathcal{S}$ that are not shared by two different fundamental triangles of $\mathcal{S}$.

To this end, put $C_1 = C$ and let S_1^1 be an arbitrary fundamental triangle with $S_1^1 \cap C_1 \neq \emptyset$. For every $k \in \mathbb{Z}$, $1 \leq k$, if $S_1^1, S_2^1, ..., S_k^1$ are defined, let S_{k+1}^1 be a fundamental triangle with the following properties: $S_{k+1}^1 \cap C_1 \neq \emptyset$, S_{k+1}^1 has a side in common with S_k^1 which is not a subset of C_1 and $S_{k+1}^1 \neq S_i^1$ for all i, $1 \leq i \leq k$. Clearly, there will always be a (smallest) number $k \geq 1$ for which no such fundamental triangle S_{k+1}^1 exists. Denoting by k_1 this number, we have defined a sequence $(S_1^1, S_2^1, ..., S_{k_1}^1)$ of fundamental triangles. Let C_2 be the union of those sides of fundamental triangles in $(S_1^1, S_2^1, ..., S_{k_1}^1)$ that are disjoint from

C_1 and not shared by two different fundamental triangles in $(S_1^1, S_2^1, ..., S_{k_1}^1)$. If $C_2 \neq \emptyset$, we construct a sequence $(S_1^2, S_2^2, ..., S_{k_2}^2)$ of fundamental triangles in a way similar to the one used for constructing of $(S_1^1, S_2^1, ..., S_{k_1}^1)$ by taking C_2 instead of C_1 (and obtaining k_2 in much the same way as we did k_1). Repeating this construction, we get sequences $(S_1^3, S_2^3, ..., S_{k_3}^3)$, $(S_1^4, S_2^4, ..., S_{k_4}^1)$, etc. We put $\mathcal{S} = (S_1^1, S_2^1, ..., S_{k_1}^1, S_1^2, S_2^2, ..., S_{k_2}^2, S_1^3, S_2^3, ..., S_{k_3}^3, ...)$ if $C_i \neq \emptyset$ for all $i \geq 1$ and $\mathcal{S} = (S_1^1, S_2^1, ..., S_{k_1}^1, S_1^2, S_2^2, ..., S_{k_2}^2, ..., S_1^l, S_2^l, ..., S_{k_l}^l)$ if $C_i \neq \emptyset$ for all i with $1 \leq i \leq l$ and $C_i = \emptyset$ for $i = l+1$.

Further, let $S_1' = T(z)$ be a fundamental triangle such that $z \notin S$ whenever S is a member of $\mathcal{S}$. Having defined S_1', let $\mathcal{S}' = (S_1', S_2', ...)$ be a sequence of fundamental triangles defined analogously to $\mathcal{S}$ (by taking S_1' instead of S_1^1). Then, one of the sequences $\mathcal{S}$, $\mathcal{S}'$ is finite and the other is infinite. Indeed, $\mathcal{S}$ is finite (infinite) if and only if its first member equals such a fundamental triangle $T(z)$ for which $z = (k, l) \in \mathbb{Z}^2$ has the property that the cardinality of the set $\{(x, l) \in \mathbb{Z}^2; \ x > k\} \cap C$ is odd (even). The same is true for $\mathcal{S}'$. If we put $\{\mathcal{S}_F, \mathcal{S}_I\} = \{\mathcal{S}, \mathcal{S}'\}$ where $\mathcal{S}_F$ is finite and $\mathcal{S}_I$ is infinite, then the conditions (a) and (b) are clearly satisfied.

Given a circle C in the graph $(\mathbb{Z}^2, A)$ which does not turn at any point $(4k+2, 4l+2)$, $k, l \in \mathbb{Z}$, let S_F and S_I denote the union of all members of $\mathcal{S}_F$ and $\mathcal{S}_I$, respectively. Then, $S_F \cup S_I = \mathbb{Z}^2$ and $S_F \cap S_I = C$. Let $\mathcal{S}_F^*$ and $\mathcal{S}_I^*$ be the sequences obtained from $\mathcal{S}_F$ and $\mathcal{S}_I$ by subtracting C from each member of $\mathcal{S}_F$ and $\mathcal{S}_I$, respectively. Let S_F^* and S_I^* denote the union of all members of $\mathcal{S}_F^*$ and $\mathcal{S}_I^*$, respectively. Then, S_F^* and S_I^* are connected by (1) and (2) and it is clear that $S_F^* = S_F - C$ and $S_I^* = S_I - C$. So, S_F^* and S_I^* are $\mathcal{B}$-components of $\mathbb{Z}^2 - C$ by (3) ($S_F - C$ is called the *inside* component and $S_I - C$ is called the *outside* component). We have proved that every cycle in the graph shown in Fig. 5 that does not turn at any point $(4k+2, 4l+2)$, $k, l \in \mathbb{Z}$, is a $\mathcal{B}$-Jordan curve.

Example 1. Consider the set of points of $\mathbb{Z}^2$ demonstrated by Fig. 6, which represents the (border of) letter K. This set is a circle in the graph $(\mathbb{Z}^2, A)$ that turns only at some of the vertices $(2k(n-1), 2l(n-1))$, $k, l \in \mathbb{Z}$, so that it is a $\mathcal{B}$-Jordan curve by Theorem 2. But, since the circle turns, at each of the four bold points, at the acute angle $\frac{\pi}{4}$, it is not a digital Jordan curve in the Khalimsky graph. For the circle to be a Jordan curve in the Khalimsky graph, it is necessary to remove, along with the four bold points, the four encircled points (because, otherwise, the circle would not even be a simple closed curve in the Khalimsky graph). But this would lead to a noticeable deformation of the image (note that the points represent centers of pixels) if the resolution of the computer screen used is not sufficiently high. This may be the case of some industrial monitors or displays.

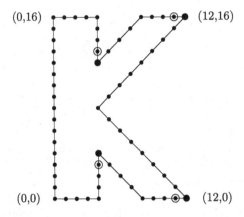

Fig. 6. A Jordan curve in $(\mathbb{Z}^2, u_3^2)$.

Remark 1. If we do not insist on structuring the digital plane by the 8-adjacency graph but admit structuring it by a factor of the graph, we may find a graph G with the vertex set $\mathbb{Z}^2$ having the property that every circle in the graph $(\mathbb{Z}^2, A)$, not only a cycle that does not turn at any point $(4k + 2, 4l + 2)$, $k, l \in \mathbb{Z}$, is a Jordan curve in G (with respect to the natural connectedness in the graph G). Let us call graphs G with this property *sd-graphs*. The *sd*-graphs are studied in [18] where it is shown that the graph demonstrated in Fig. 7 is a minimal (with respect to the set of edges) *sd*-graph. Note that this graph is even a factor of the Khalimsky graph.

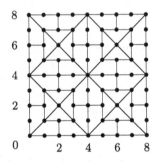

Fig. 7. A portion of a minimal *sd*-graph.

4   Conclusions

We have found a structure on the digital plane $\mathbb{Z}^2$, the graph $(\mathbb{Z}^2, A_8)$ together with the set $\mathcal{B}$ of paths of length 2, which provides the plane with a connectedness allowing for a digital analogue of the Jordan curve theorem (Theorem 2).

This means that the graph $(\mathbb{Z}^2, A_8)$ together with the set $\mathcal{B}$ may be used as a background structure on the digital plane for the study and processing of digital images. An advantage of the $\mathcal{B}$-Jordan curves in the graph $(\mathbb{Z}^2, A)$ over the Jordan curves in then Khalimsky plane is that they may turn, at some points, at the acute angle $\frac{\pi}{4}$. Hence, the graph $(\mathbb{Z}^2, A_8)$ endowed with the set $\mathcal{B}$ provides a variety of Jordan curves richer than the one provided by the Khalimsky topology. Thus, the graph offers a convenient alternative to the topology. Since Jordan curves represent borders of objects in digital images, the structure on $\mathbb{Z}^2$ given by the graph $(\mathbb{Z}^2, A_8)$ with the set $\mathcal{B}$ may be used in digital image processing for solving problems related to boundaries such as pattern recognition, boundary detection, contour filling, data compression, etc.

Acknowledgement. This work was supported by the Ministry of Education, Youth and Sports of the Czech Republic from the National Programme of Sustainability (NPU II) project IT4Innovations excellence in science - LQ1602.

References

1. Chvátal, V.: Correction to: a de Bruijn-Erdős theorem in graphs? In: Gera, R., Haynes, T.W., Hedetniemi, S.T. (eds.) Graph Theory. PBM, pp. C1–C2. Springer, Cham (2018). https://doi.org/10.1007/978-3-319-97686-0_15
2. Engelking, R.: General Topology. Państwowe Wydawnictwo Naukowe, Warszawa (1977)
3. Khalimsky, E.D., Kopperman, R., Meyer, P.R.: Computer graphics and connected topologies on finite ordered sets. Topology Appl. **36**, 1–17 (1990)
4. Khalimsky, E.D., Kopperman, R., Meyer, P.R.: Boundaries in digital plane. J. Appl. Math. Stochast. Anal. **3**, 27–55 (1990)
5. Kiselman, C.O.: Digital jordan curve theorems. In: Borgefors, G., Nyström, I., di Baja, G.S. (eds.) DGCI 2000. LNCS, vol. 1953, pp. 46–56. Springer, Heidelberg (2000). https://doi.org/10.1007/3-540-44438-6_5
6. Kong, T.Y., Kopperman, R., Meyer, P.: A topological approach to digital topology. Amer. Math. Monthly **98**, 902–917 (1991)
7. Kong, T.Y., Roscoe, W.: A theory of binary digital pictures. Comput. Vision Graphics Image Proc. **32**, 221–243 (1985)
8. Kong, T.Y., Rosenfeld, A.: Digital topology: introduction and survey. Comput. Vision Graphics Image Proc. **48**, 357–393 (1989)
9. Kopperman, R., Meyer, P.R., Wilson, R.G.: A Jordan surface theorem for three-dimensional digital space. Discrete Comput. Geom. **6**, 155–161 (1991)
10. Melin, E.: Digital surfaces and boundaries in Khalimsky spaces. J. Math. Imaging Vision **28**, 169–177 (2007)
11. Melin, E.: Continuous digitization in Khalimsky spaces. J. Approx. Theory **150**, 96–116 (2008)
12. Rosenfeld, A.: Connectivity in digital pictures. J. Assoc. Comput. Math. **17**, 146–160 (1970)
13. Rosenfeld, A.: Digital topology. Amer. Math. Monthly **86**, 621–630 (1979)
14. Rosenfeld, A.: Picture Languages. Academic Press, New York (1979)

15. Šlapal, J.: Jordan curve theorems with respect to certain pretopologies on $\mathbb{Z}^2$. In: Brlek, S., Reutenauer, C., Provençal, X. (eds.) DGCI 2009. LNCS, vol. 5810, pp. 252–262. Springer, Heidelberg (2009). https://doi.org/10.1007/978-3-642-04397-0_22
16. Slapal, J.: A jordan curve theorem in the digital plane. In: Aggarwal, J.K., Barneva, R.P., Brimkov, V.E., Koroutchev, K.N., Korutcheva, E.R. (eds.) IWCIA 2011. LNCS, vol. 6636, pp. 120–131. Springer, Heidelberg (2011). https://doi.org/10.1007/978-3-642-21073-0_13
17. Šlapal, J.: Graphs with a path partition for structuring digital spaces. Inform. Sci. **233**, 305–312 (2013)
18. Šlapal, J.: Convenient adjacencies on $\mathbb{Z}^2$. Filomat **28**, 305–312 (2014)

Author Index

Printed in the United States
By Bookmasters